Marieme Fall

Salinization and its consequences on market garden crops

Marieme Fall

Salinization and its consequences on market garden crops

in the Niayes (Senegal's northern coast)

ScienciaScripts

Imprint

Cover image: www.ingimage.com

This book is a translation from the original published under ISBN 978-620-6-71324-1.

Publisher:
Sciencia Scripts
is a trademark of
Dodo Books Indian Ocean Ltd. and OmniScriptum S.R.L publishing group

120 High Road, East Finchley, London, N2 9ED, United Kingdom
Str. Armeneasca 28/1, office 1, Chisinau MD-2012, Republic of Moldova, Europe
Managing Directors: Ieva Konstantinova, Victoria Ursu
info@omniscriptum.com

Printed at: see last page
ISBN: 978-620-8-57069-9

Contents

"Science and technology, within the framework of their contribution to economic and social development, should be used to detect, avoid or limit the dangers threatening the environment and to solve the problems posed by it, and in general for the good of mankind". (Principle 18 of international environmental law: basic texts and references), Dommen and Cullet, 1998, page 9.

Dedication

To my father **Seyni Maissa Fall**
To my mother **Khady Fall**
To my grandparents **Allé Birima Fall** and **Sakha Fall**, may the Almighty welcome them to his eternal home.
To my family, **brothers** and **sisters**
To my husband
To all those who have constantly given me advice and wished me success.

Acknowledgements

We give thanks to **God** the Almighty, the Clement, the Merciful, for his omnipresence at our side, granting us life, health and the courage to overcome the trials of life until the writing of this thesis.

This thesis would not have been possible without the help and invaluable contributions of a large number of people.

We would like to thank our thesis supervisor, Pr Boubou Aldiouma Sy, for agreeing to supervise this work. His scientific rigour and exacting standards guided us through most of the problems, while ensuring that the theme was respected, clear and simple. This work would never have been completed without him. We are grateful to the members of our thesis committees, Dr Bodian, Dr Abou Sy and Dr Souleymane Niang, who agreed to follow the progress of this work. Their advice and guidance were invaluable.

Special mention should be made of the professors in the geography department, in particular Mr Omar Diop, Adrien Coly, Da Dieng, Cheikh Wade and Diakhaté, whose contribution to our scientific training provided the basis for this thesis. The immense human qualities found in each of them, their skills and their advice in difficult moments, helped us a great deal.

We would like to thank all those who contributed to the preparation of this work, and those who gave us material and/or moral support. Sincere thanks to the head of the SAED laboratory in Ross Béthio, Mr Mouhamadou Diop, and the head of the laboratory in the geology department at UCAD, Mr Moussa Sow, for respectively supervising and carrying out the analyses of soil and water samples in 2010 and 2018, as part of this study. Their scientific advice on reading the results was a great help.

Not forgetting those who made our task easier during the data collection phase.

Our contacts, Mr Siré Ndiaye, in Malika, Mr Ousmane Ndiaye, Mr Ngaty Diop and Mr Dieumb Diop, in Taré, Mr Dické Diop in Darou Mboumbaye, Mr Ngagne Niang and Mr Lamine Dieng, in Diender, who contributed a great deal to the data collection in the field, through sampling and interview guides. Mr Mbacké Guéye, my dear husband, for accompanying us to take soil and water samples at Lake Wouye in Malika in 2010, not forgetting his financial and moral support. Mr Abdoulaye Fall, agent of Enda graf Thiès, Mr Faye agent of the DGPRE, Mr Ka agent of the Millenium village in Louga, M.

Pape Niang from the Gueumbeul reserve, who respectively helped us find a site in Thiès, draw up certain maps and choose the Taré and Ndiébène Gandiol sites.

Many thanks to the headmaster of Demba Diop high school in Mbour until 2015, Mr Robert Pouye, for his understanding, moral and administrative support, which enabled us to combine studies and work. Thank you to all the market gardeners in the different areas we studied, for their availability in answering our questions, thanks to them we were able to carry out this study.

Foreword

The area covered by the Niayes is changing: silting up, salinisation of the water, alkalinisation of the land, wind erosion and heavy urbanisation. The Niayes provide almost two-thirds of Senegal's market garden produce, and salinisation of the water table is a recurring phenomenon that poses huge problems for stakeholders. The analytical results of our 2007 DEA (Diplôme d'Études Approfondies) dissertation on the salinisation of the soils of Lake Wouye in the district of Malika: impacts on the natural environment and socio-economic activities, showed a relatively high salt content in the lake due to the phenomenon of saline intrusion resulting from the proximity of the Atlantic Ocean. This salt intrusion is having a negative impact on market gardening.

We wanted to take our research further by looking at the whole Niayes region, to see whether the problem is general or localised. The theme of this thesis is salinisation in the Niayes and its impact on market gardening. Market gardening in the Niayes is threatened if no natural resource management policy is applied to safeguard and restore the market garden basins. Watering plants with saline water has a negative impact on the soil and on market gardening. A lot of land has been abandoned due to productivity problems caused by the use of salt water for irrigation.

This is the background to this research on the Niayes, which stretch from the Dakar region to Saint Louis. This research is a contribution to better management of natural areas, and to taking into account the concerns of the people living in the area.

The aim of this work is to show the saline dynamics of the Niayes and its consequences on market gardening.

This thesis is being carried out at the Leüdi 'Dynamiques des territoires et développement' laboratory of the Doctoral School of Human and Social Sciences (ED-SHS), UFR de Lettres et Sciences Humaines de l'Université Gaston Berger (UGB) de Saint-Louis, *Ecosystems and Environment* course, *Geomorphology, Environment and Development* (GED) team.

The data consists of bibliographical sources, *in situ* measurements and laboratory results. Relevant data from monitoring results from 2010 to 2018, enabling the salinisation processes to be characterised, are presented. The data shed light on the problem of salinisation in the Niayes and its impact on market gardening.

Introduction

The aim of this research is to gain a better understanding of the degradation of irrigation water by salinisation, in the current context in Senegal. For more than three decades, we have been witnessing the

In Senegal, there has been a steady deterioration in the soil, resulting in a steady decline in agricultural performance. The real problem facing Senegalese agriculture remains the pronounced loss of soil fertility and the quality of irrigation water. Studies carried out by the CILSS in November 2010 show that of the 3,805,000 ha of arable land available to Senegal, some 2,400,000 ha are severely degraded, or 63%. This soil loss is mainly due to the combined effects factors such as the use of chemical fertilisers, slash-and-burn cultivation, massive deforestation, overloading with livestock, poor quality of irrigation water, and above all the decline and instability of rainfall, partly as a result of the decline in natural woodland, with some 20,000 ha of forest lost every year (Enda Pronat, 2018).

This thesis is a scientific contribution to the evolution of the *Niayes* and the state of degradation of irrigation water and arable land, with particular emphasis on the results already obtained in this area. The aim of this research, which is divided into five chapters, is to show that salinity remains a recurring problem that is detrimental to market gardening activities.

Chapter 1 describes the physical setting of the *Niayes*, and more specifically the *Niayes* of Dakar, Thiès, Louga and Saint-Louis. It presents market gardening activities and salinisation factors.

The second chapter describes the literature review, the theoretical foundations of our research, the conceptual framework and the operationalisation of the study concepts.

The third chapter is devoted to the methodological approach, to show the various stages followed in this study in collecting data: taking soil and water samples for laboratory analysis.

The fourth chapter presents and describes the results of analyses and surveys, and even the situation and evolution of salinisation in the *Niayes*. It is based on our various analysis results, to find out the pH and electrical conductivity of our soil and water samples in 2010 and the chemical component of irrigation water in 2018.

The fifth chapter analyses, interprets and discusses the results obtained from the sample analyses and surveys. Emphasis will be placed on the causes and consequences of the phenomenon of salinisation in the *Niayes*, the natural and anthropogenic causes and their impact on market gardening activities. This section shows a decline in arable land, market gardening activities and yields in the *Niayes*.

Chapter 1

The Niayes: physical environment, market garden crops and salinisation factors

In Sahelian countries, soil degradation has been underway for several decades. The phenomenon was already reported in Senegal during colonisation (Hubert, 1917: 377): a "drying out in the Senegalese regions". Aubert, 1946 predicted soil degradation in the Louga region. Kane (1985) pointed out that forest retreat is associated with soil fragility. Since independence in 1960, the degradation of cultivated land has taken on worrying proportions, to the extent that in 1993 the Economic Commission for Africa predicted an increase in poverty, overpopulation and famine at a high level in countries such as Senegal by 2008.

At present, the harmful effects of soil degradation in Senegal extend far beyond the rural world and affect the entire national economy. This overall degradation, which results from water and wind erosion, salinisation and acidification, has been estimated at a minimum of 1% of GDP per year. This means that every year, the country loses at least 85 to 90 billion CFA francs (Dièye, 2016). The country's socio-economic development is largely based on agriculture, which makes a significant contribution to GDP, amounting to 7.0% in 2012, 6.8% in 2013 and 15.4% in 2017, with an increase around 6% in 2018, compared with 22.1% in 1960 (CSE, 2015; World Bank Group, 2018).

A study published in 2016 by the ANSD on *the economic and social situation in Senegal in 2013* shows a fall in agricultural yields in the 2013/2014 season, which did not match the performances seen in 2008, 2009 and 2010. Speculative crops (cereals and oilseeds) recorded declines. For example, agricultural production fell in 2014, mainly due to a 5% drop in output and low yields for several agricultural products, such as groundnuts (739 kg/ha in 2013, compared with 977 kg/ha in 2012), sesame (3,689 tonnes in 2013, down 30.3%), industrial tomatoes (-13.0%, with 80,000 tonnes), cassava (-12.8%), cowpeas (-26.0%, with 40,688 tonnes in 2013, compared with 55,015 tonnes in 2012). A decline in yields has been noted for several vegetable products, with the exception of onions, which have seen an increase since the 2000s, with 230,000 tonnes in 2013, compared with 210,000 in 2012. The 2017-2018 crop year saw good onion production, with more than 400,000 tonnes (APS, 2018). Despite the continuing fall in agricultural yields, Leiseinger et al (1995) consider that agriculture remains the mainstay of the Senegalese economy. It has been set up as the country's growth engine since adoption of the Emerging Senegal Plan (PSE) in 2014.

The many studies carried out on this phenomenon (Gavaud, 1990), Charollais and Weber (1994), Schiess (1995), Ndour (2001), Diallo et *al* (2015), have used different approaches at various scales. The few available data obtained to date attest to the complexity of the phenomenon, which continues to be a very active research topic with as many unanswered questions as explicitly provided answers on soil degradation, poor quality irrigation water, etc.

1. The context of the study

In wetlands such as the *Niayes* in Senegal, the shallowness of the water table encourages significant capillary rise in depressions during the dry season, which, due the high evapotranspiration potential, concentrates on the spot and becomes sodic and alkaline. Salinisation and acidification are among the main factors leading to land degradation, a reduction in arable land and lower yields. Acidification affects fluvio-marine soils in the

Senegal River Valley (400,000 ha), Casamance (around 400,000 ha), Sine Saloum (230,000 ha) and the *Niayes* (5 to 8,000 ha) (Sadio, 1991). Salinisation is a particularly serious problem for the country, with negative repercussions on agricultural activities such as market gardening. Figure 1 shows the estimated salinisation of arable land in Senegal, by agro-ecological zone.

Figure 1: Estimated salinisation of land in Senegal (in ha)

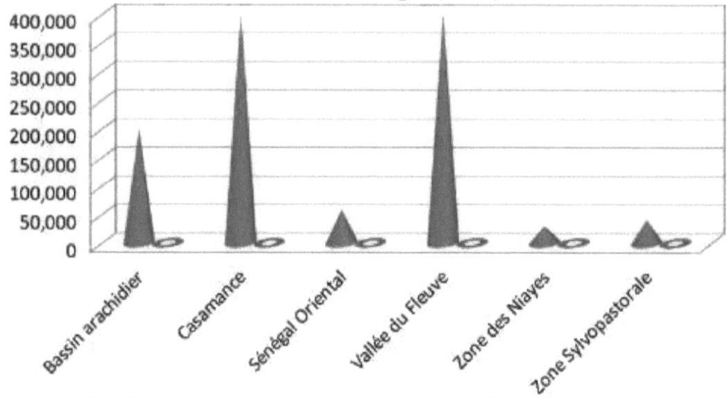

Source: Data from the National Institute of Pedology (INP, 2013)

An estimate of the salinisation of arable land in Senegal shows the vulnerability of the country's soil resources, through both natural and man-made causes (human activities). The soils are less and less suitable for production, with increasingly serious consequences for yields. Soil degradation through salinisation is compromising agricultural activities, particularly market gardening in the Niayes agro-ecological region. The phenomenon is on the increase, despite the fact that a large proportion of the country's vegetable production comes from the *Niayes.*

Figure 2: Location of the *Niayes*

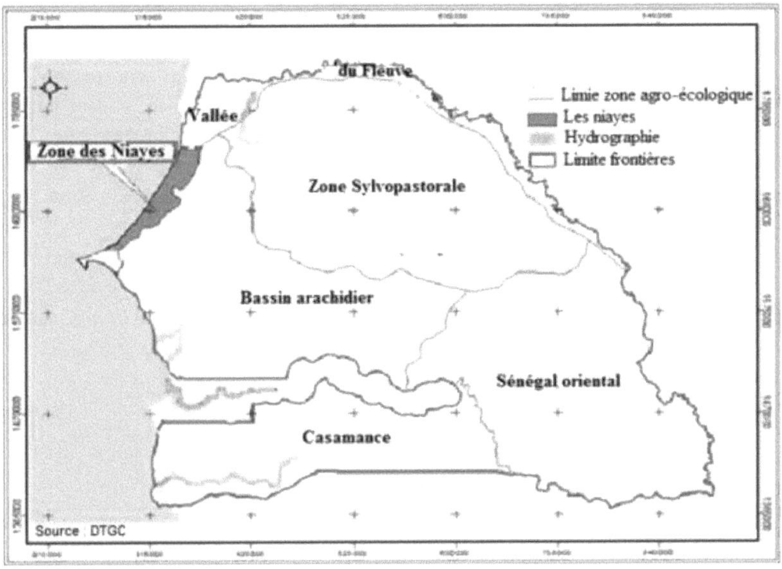

The area covered by the *Niayes* is one of the most agriculturally fertile in Senegal. The *Niayes* occupy the northern coast of Senegal, stretching from Gandiolais to the Cap Vert peninsula (in the department of Pikine). The area lies between 14°34'23" N and 16°05'02" N, and between 16°01'59" W and 17°31'46" W (Diouf, 2013), over more than 180 km in length and 30 to 35 km in width (Aguiar, 2009), with a surface area of around 3,090 km^2 (Sy, 2006). It forms the western edge of the Sahel. Belonging to the Canary Islands climatic zone, its proximity to the Atlantic Ocean favours a relatively low average annual temperature: 24°C to 28°C. The rainfall pattern is characterised by a rainy season running from July-September to October, with annual cumulative rainfall of around 500 mm in the south and 300 mm in the north in 2009. Since 2010, there has been a delay in rainfall and a retreat of the isohyets towards the south of the country (Figure 3). It sometimes arrives late (August-September) in some areas (Thiès), and sometimes ends before October due to the rapid retreat of the ITD. Annual rainfall totals almost never reach 300 mm in the north and 500 mm in the south (around 252 mm in Saint-Louis and 350 mm in Dakar in 2012) (météo Sénégal, 2012). The *Niayes* benefit from a nuanced climate due to the maritime trade winds and ocean currents. This attenuates the seasonal thermal contrasts of the Sahelian zonal climate, which becomes less pronounced the further you move away from the coast. The use of the *Niayes* for market gardening is increasing all the time, hence the strong pressure it is under. The depth of the water table is highly variable. The water level in wells and "céanes" (traditional wells dug in areas where the water table is not deep) is generally between 2 and 10 metres[1] (in depressions), while for wells located further inland (at the top of the dunes), the water table is between 15 and 30 metres .[2]

[1] Field surveys and in situ measurements
[2] Field surveys and in situ measurements

This favourable context has made the Niayes the site of intense agricultural activity in the field of market gardening. Since the early 1970s, levels have been falling steadily due to fluctuations of varying severity. Over the last two decades, low rainfall has resulted in a general drop in the water table. Reduced rainfall led to a gradual fall in the water table, averaging 0.51 m/10 years between 1958 and 1994. In 2002, the water table appears to have continued to fall despite the stabilisation of the fall in rainfall since the end of the 1990s (1998, 1999 and 2000). The level of the water table, which is directly influenced by rainfall, is subject to a delayed effect, known as the "memory effect", which occurs every few years (Aguiar, 2010). It has fallen more in the southern sector (where the Quaternary Sand Table is heavily exploited) than in the north.

In this context, evapotranspiration, which accounts for almost 90% of outflows and 93% of natural losses, is the main factor in groundwater discharge (Aguiar, 2009). Figure 3 shows changes in cumulative annual rainfall from 2009 to 2018 in Dakar, Thiès, Louga and Saint-Louis.

Figure 3: Annual rainfall totals from 2009 to 2018

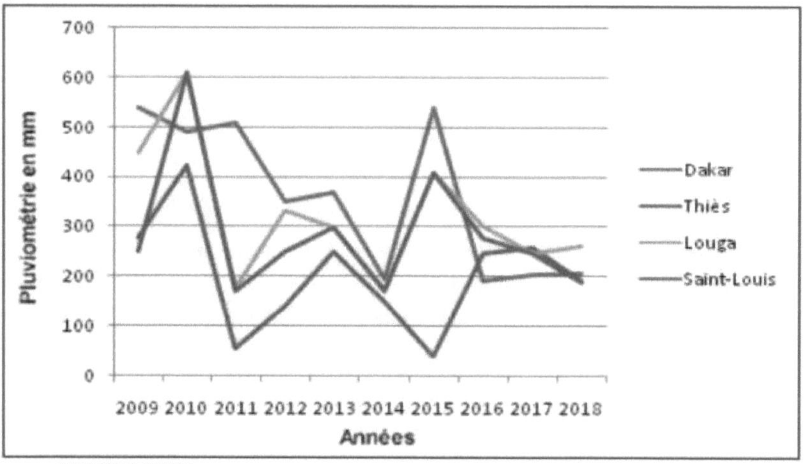

Source: ANACIM (2018)

Since 2010 (a wet year in all four regions), rainfall has been irregular. Generally speaking, there has been irregularity over the years, with 2011 and 2014 being the lowest (less than 200 mm in the north and south in 2014). 2018 was not a rainy year, with less than 500 mm in the south and less than 300 mm in the north (around 205 mm in Dakar and 190 mm in Saint-Louis) ANACIM (2018). The low level of water replenishment means that there is a risk saline intrusion in the *Niayes* located by the sea. An increase in water salinity could compromise market gardening and the survival of the *Niayes*. The salinisation of the *Niayes* area has several causes.

It is mainly of ancient or current marine origin and is characterised by seepage from the sea into an initially soft coastal water table (Sanokho, 2007). Ancient marine salinisation is linked to the Nouakchottian marine transgression: 12,000 - 2,000 years *BP* (Sow 2001), which left areas of "trapped water" that became lagoons and Nouakchottian gulfs (Ba et al, 2005).

The *Niayes* is the country's main market-gardening area, among the other sites (such as the river valley). At present, saline intrusion into the water table and acidification due to insufficient leaching are being observed. According to the Institut National de Pédologie (INP, 2013), this has led to the salinisation around 5,000 ha of arable land in the *Niayes*.

2. Salinisation factors in the *Niayes*

2.1. *Niayes* soils and salinisation

From north to south, the soils of the *Niayes* are quite diverse, with a surface horizon of sandy-silty texture with permeability varying from 20 to 150 cm/h (Ndoye, Ndiaye and Diop, 2006). They are varied, but the most common are non-leached tropical ferruginous soils or *dior* soils, poorly developed sandy soils, organic hydromorphic soils and humus-bearing clayey swamp soils. They are scattered in different places depending on the topography.

Figure 4: Structural section of the *Niayes* region

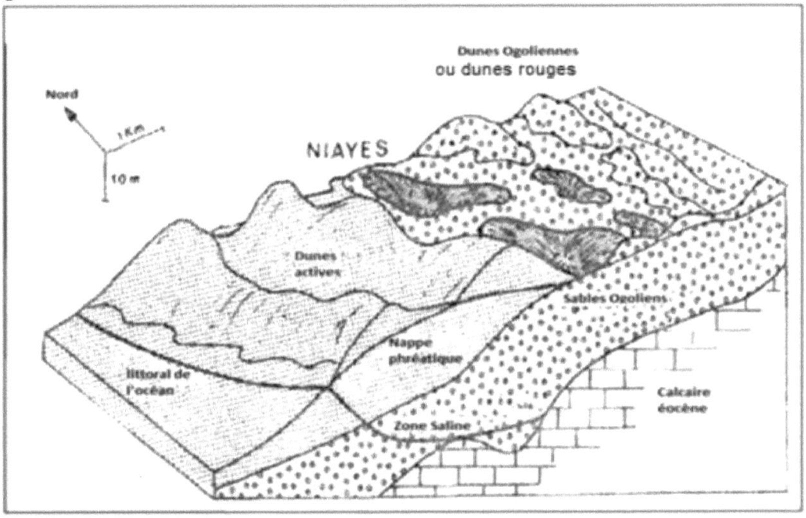

Source: BRGM 1984, Lazard 2009. Image modified by M. Fall, 2015

The red dunes are dominated by tropical ferruginous soils with little or no leaching, known as "*Diors*" soils, which are slightly acidic and occupy most of the *Niayes* region, with a very low organic matter content of less than 1%. They are subject to alkalinisation, wind erosion and run-off. On the slopes of the dunes, leached tropical ferruginous soils appear. These soils are no richer than their predecessors, but they are facing a problem of degradation through salinisation and alkalinisation, which has been increasing over the years as global warming increases evapotranspiration and saline upwelling. Towards the edges of the depressions, humus-bearing, red-brown or isohumus soils with varying degrees of marshiness, known as "Diors noirs" or "Bambo", develop in the north-western part of Louga and south-eastern part of Saint-Louis; these soils support market gardening very well. Their weak point is the fairly rapid loss of organic matter when land is cleared. Strictly speaking, the *Niayes* are inter-dune depressions, providing the transition between the semi-fixed yellow dunes and the Ogolian continental system, on the one hand, and between the white cordon adjacent to the beach and the semi-fixed yellow dunes (Ndioukis), on the other. *In addition to* these depressions with organic hydromorphic soils, the market gardeners of the northern coastline incorporate ancient fossil valleys in the Niayes around the Dakar and Thiès regions.

Soil types include grey-brown to blackish hydromorphic soils with temporary or permanent waterlogging, which are very decisive in the depressions of the *Niayes*, and are of great interest for vegetable production.

Soil salinisation is more marked in these low-lying areas (*Niayes*) where the water table is shallow and subject to marine intrusion. In the sharp dunes (sharp coastline), there are raw mineral soils brought in by the wind, which are characterised by their poverty or lack humus-bearing horizons: non-climatic raw mineral soil.

Vertisols are found in the Sébikotane sector (Bargny plateau) and along the Somone-lac Tanma axis, and are often formed in regions where basic rocks predominate. They are found mainly at the foot of the plateau in the form of colluvial glacis or in the shallows and basins of the valleys, covered stagnant water during the rainy season. They often have a pH close to neutral and are low in organic matter (OM). Halomorphic soils persist in the vicinity of coastal lagoons blocked dune belts in the southern part of the *Niayes* (Dakar and Thiès regions. The salts come from the saline upwelling in the estuaries during low-water periods, but are also partly inherited from the lacunar regime at the end of the last Nouakchott transgression and incorporated into the sediments. Halomorphic soils extend to the edges of permanent or temporary salt lakes, such as Lake Tanma. The *Niayes* area has a wide variety soil types, with pH levels varying according to the environment.

In terms of salinity, this research shows the existence of four soil entities: extremely saline soils, in the Niayes of Thiès, slightly saline to saline soils, in the lowlands of the *Niayes* of Pikine in Dakar and non-saline soils, in the *Niayes* of Thiès and Saint-Louis more or less distant the sea and Lake Tanma. There is also a very strong presence very alkaline soils throughout the *Niayes* area. This situation is likely have an impact on vegetation, and hence on vegetable production.

2.2. *Niayes* vegetation and salinisation

In the *Niayes*, fluctuations in salt content are responsible for the differentiation of two types of vegetation, one linked to the salty substrate and the other to the freshwater table. When the balance between freshwater and saltwater in the subsoil is disrupted, or as a result of marine intrusion, salinisation of the water table and soil can occur (Diallo et *al.*, 2009). In these saline environments, species such as *Blutaparon vermiculare, Paspalum vaginatum, Blumea viscosa and Diplachne fusca* are pioneers and are found in the Grande *Niaye* de Pikine.

When environmental conditions become unfavourable, the vegetation evolves towards vegetation that is less demanding in terms of water and salt, leading to the establishment of more or less indifferent therophytes. In deep depressions that are permanently flooded, hygrophilous or mesohydrophilous vegetation develops, consisting of *Typha domingensis* (as in Lake Wouye in 2018), *Phragmites australis* and *Paspalidium geminatum*. Salinisation of the *Niayes* area has led to the disappearance of many plant species (disappearance of *Echinochloa sp.* groups), PNEEB (2014). The vegetation is made up of species indicative of degraded soils, such as *Eragrostis*, *Opuntia tuna* and *Typha australis*. Present on the yellow and white dunes of *the Niayes* area, the *Cactus* or *Opuntia* tuna adapts to sensitive areas (alkalinity, salinisation) and often proliferates very quickly. A strong presence of the *Cactus* species has been noted, especially in the villages of Taré and Darou Mboumbaye (favoured by anthropochory and reinforced by land degradation). One study showed that: "the growth of cacti planted in salty soils was not significantly affected by different salt concentrations, and that most were stimulated by the increase in salt concentration in the environment" (Khales and Baaziz, 2006).

Photo 1: *Cactus* Taré and Darou Mboumbaye in 2010 and 2018

Source: Photo M. Fall, 2010 and 2018

The intensive development of market gardening throughout the *Niayes* area is contributing to soil impoverishment and vegetation degradation. Added to this is the over-exploitation of the depressions which, together with the region's rainfall deficit, has led to salinisation in places, and the disappearance of several plant species, such as *Nymphaea lotus*, a more water-demanding aquaphilous species found in the centre of the depression (CRDI, 2001). Winds are an essential element in monitoring changes in the salinisation of the *Niayes*.

2.3. Wind and salinisation of the *Niayes*

The maritime trade winds that blow over the *Niayes* contribute to erosion and salinisation. These winds encourage the removal of topsoil and bury the market garden basins in sea sand that is unsuitable for cultivation because of its salinity (Ndao, 2012). Wind speeds are lower during the rainy season, in July, August and September, at 21 and 26 km/h, and very high during the dry season, at 24 and 30 km/h, causing sea spray to bury market gardening areas (especially in the Ndioukis), leading to alkalinisation and salinisation.

Figure 5: Wind speeds in 2018 in the *Niayes* area/km/h

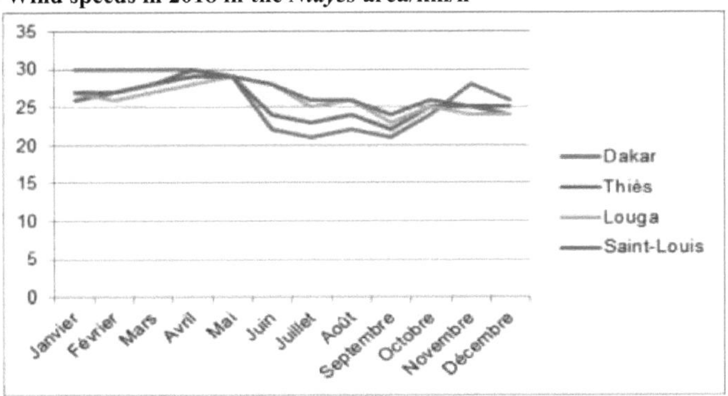

Source: weather data (2018)

The coastal strip's exposure to the maritime trade winds is largely responsible for the originality of the environment, but also for the intrusion of sea sand (sea spray), which encourages alkalinisation and salinisation in the area. The maritime trade winds are capable of having an effect up to 200 km inland, promoting the advance of sea spray towards the market gardening areas in the continental part of the *Niayes* region. From May onwards, the monsoon regime begins with a strengthening of the southerly winds, to the detriment of the northerly winds. The harmattan is felt on the coast for a short time, but contributes to a significant rise in temperature, which accentuates evapotranspiratory demand, and therefore saline rise.

2.4. Temperatures and salinisation

Temperatures recorded in the area covered by the *Niayes* vary from 15°C to 35°C. They never reach 40°C and are conducive to water conservation over a long period, which favours market gardening. They are recorded 9 months out of 12 throughout the region. The *Niayes* are reputed to have ideal temperatures for the development of market gardening, but with global warming in recent years, temperatures have risen (26°C to 35°C), leading to heavy evaporation and rising salinity, especially in the Thiès and Louga regions. There has been a decline in rainfall from north to south and an increase in evaporation. Heavy evaporation is causing groundwater to dry up, and some rivers to dry out, making it difficult to grow vegetables. Evaporation alkalinisation and salinisation in the *Niayes*.

2.5. Precipitation and salinisation

In the Niayes region, rainfall used to be recorded from July to October, but as a result of climatic disturbances, it now varies from July to September. Sometimes, the first rains arrive in August. Recorded rainfall varies between 500 mm in the south and 300 mm in the north (ANACIM, 2016). The rest of the year (October to July) is characterised by a virtual absence of rain. The few rains recorded barely exceed 50 mm in total. These rains are known as "out-of-season rains" or "heug". This drop in rainfall in the *Niayes* has negative repercussions on the environment, with a drop in the water table, which encourages a rise in salt levels to the surface, causing alkalinisation and salinisation of the soil. Figure 6 rainfall trends in *the Niayes* region from 2010 to 2016.

Figure 6: Rainfall trends in *the Niayes* region from 2010 to 2016

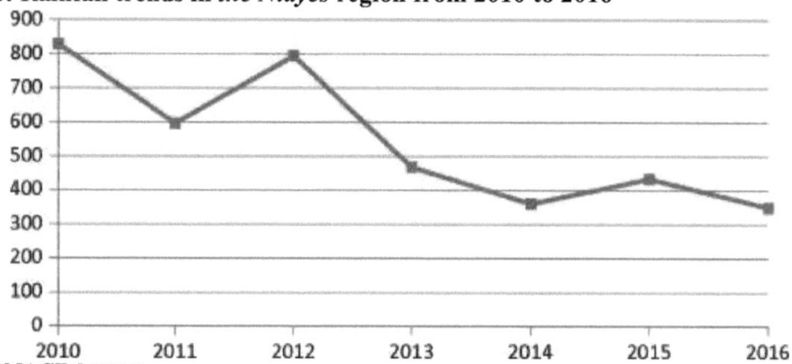

Source: ANACIM (2016)

From 2010 to 2016, cumulative rainfall varied from 350 to 829 mm. The lowest level was recorded in 2016, with low rainfall, and the highest was in 2010, with a good average. In 2016, rainfall varied between 150 and 350 mm in the *Niayes* region. Figure 7 trends in annual cumulative rainfall from 2012 and 2015 in Dakar, Thiès, Louga and Saint-Louis.

Figure 7: Annual cumulative rainfall from 2012 and 2015

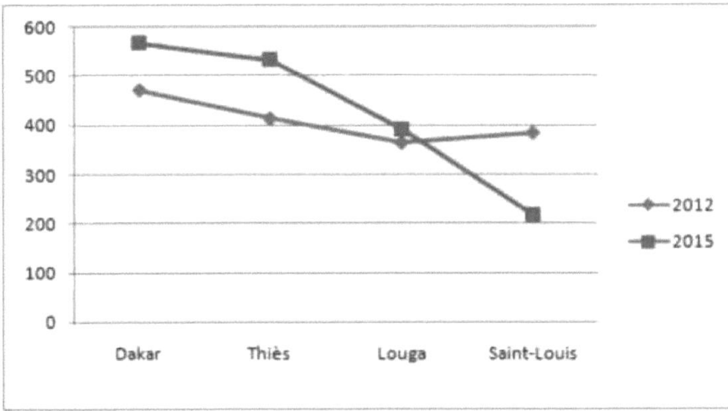

Source: ANACIM (2015)

Between 2012 and 2015, rainfall was irregular in the *Niayes* area, particularly in Saint-Louis, with 384 mm in 2012 compared with 219 mm in 2015, a difference of 165 mm. Rainfall varies from around 200 to 500 mm per year. This has an impact on groundwater recharge, leading to salinisation and difficulties in market gardening. Rainfall variations compared with the annual average at the Saint-Louis and Louga stations show that the amount of rainfall varies from one year to the next. Over the last ten years, there has been a remarkable drop in rainfall, which has had an impact on the development of market gardening. The importance of the *Niayes* area for market gardening needs to be demonstrated.

3. The *Niayes* as a support for market gardening activities

The *Niayes* are wide and long on the Cape Verde peninsula, but narrow in the Thiès region, forming numerous pools that dry up in the dry season and fill up in the winter months. They diminish the further north you go. These depressions are ideal for growing vegetables. The *Niayes* site is more concentrated in the Dakar region (Cape Verde peninsula) and in the Thiès region (Mboro), where it forms areas characterised by an almost permanent presence of water, unlike the Dior soil site, where there is greater infiltration. It provides around 60% of market garden crops: onions, potatoes, carrots, kale, cherry and table tomatoes, sweet aubergine, lettuce, chilli, bitter aubergine (jaxatu in Wolof), etc. (GRET, 2015). Market gardening takes place at two times: the cold counter-season from October to March and the hot counter-season from April to June. The crops most grown in the cold counter-season are onions, chillies, bitter aubergines (jaxatu), cabbage and sweet aubergines.

4. The *Niayes* market economy

The *Niayes* is first and foremost a specific ecological region made up of a group of dunes and depressions of different ages, textures and colours. This is why the typology borrows from the ecosystem its differentiations, which are reflected in the various production sites that can be observed in discontinuous patches in this natural environment. Around 80% of Senegal's horticultural production comes from the *Niayes* (Diop et al, 2019), which supplies fresh produce to Greater Dakar and the other regions that make it up. There are two types of horticultural farms: market garden farms and tree farms. At national level, market gardening, mainly in the *Niayes*, produced more than 106,000 tonnes in the 2017-2018 crop year (Guèye, 2018), compared with 150,000 tonnes in 2003 (ISRA, 2003). The presence of water and the diversity of vegetation justify the fertility of the land, encouraging the appearance of the first market gardeners on the sites since the drought of the 1970s, start of a long period market gardening in the *Niayes*.

5. Morphopedological units

A number of morpho-sedological units have been identified: the *Ndioukis* (low-lying areas between white dunes and semi-fixed yellow dunes), the *Niayes* (low-lying areas between semi-fixed yellow dunes and red Ogolian dunes) and the valleys or fossil lakes.

5.1. The *Niayes*

The Niayes correspond to interdunal depressions characterised hydromorphic organic soils where the water table is shallow, or even outcrops in some places, and where Guinean-type vegetation still persists. These depressions are ideal for market gardening. Given their small size, the cultivated plots are fairly small, varying from around 0.1 to 1 ha (Fall et *al*, 2001), which explains the high density of plots and the diversity of crops. Cramped conditions and salinity in places remain a handicap to the production system in the *Niayes*.

5.2. The *Ndioukis*

The *Ndioukis* are market-growing basins located between the system of living dunes and semi-fixed dunes (white dunes and yellow dunes). They are found mainly between the coastal dunes from Kayar to Gandiolais, where onion cultivation is dominant, with national production of around 367,000 tonnes in 2015, compared with 245,000 tonnes in 2014 and 90,000 tonnes in 2001. The presence of shallow freshwater duckweed has led to the development of market gardening and specialisation in carrot and turnip production. In addition, they benefit from good protection against north and north-westerly winds thanks to reforestation areas planted with *Casuarina equisetifolia* along the entire northern coastline. The *Ndiouki* production system is much more prevalent in the central *Niayes* (Thiès region) and the northern Niayes (Louga and Saint-Louis regions), and the size of farms remains small, not exceeding 1 ha (Fall et *al*, 2001). The size of farms is getting smaller and smaller, due to the major constraint that remains the rising salinity. Given the poor soils, the *Ndiouki* production system requires large inputs of mineral and organic fertilisers. The constraints are the isolation of these areas, which are difficult access, with impassable roads and sandy tracks.

Photo 2: View of a *Niayes forest*

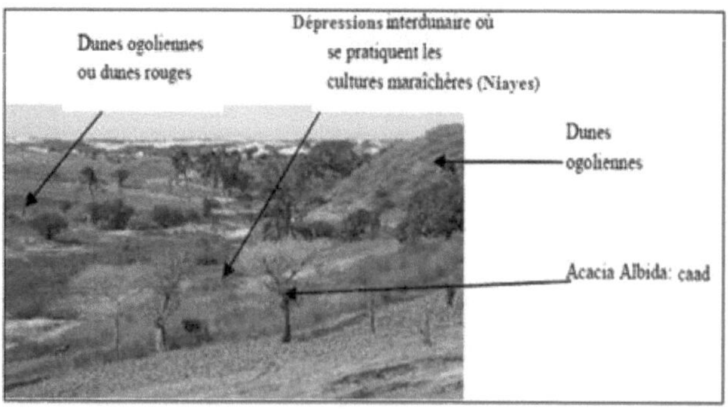

Source: A-M. Lézine & L. A. A. Aguiar (December 2004). Photo modified by M. Ndao, 2012 and M. Fall, 2018

Dry valleys or lakes

Dry valleys and lakes are former hydrographic networks that have become dry due to climatic deterioration, and are used by market gardeners. This farming system is specific to the Dakar and Thiès regions, where there is a concentration of dry valleys. The limiting factor is the salt that invades the land and the water table. The drought of the 1970s led to the drying up of the old hydrographic networks, which were more concentrated in the Dakar region (Lake Mbeubeuss, Reubeut, Malika and Yeumbeul, etc.) and in the Thiès region with Lake Mbawane: the Somone-Lac Tanma axis, etc. Once dotted with perennial lakes running parallel to the coast, the *Niayes* region has lost a significant number of them as a result of the deterioration in climatic and rainfall conditions.

At present, the minimal surface run-off that can be observed in this sector is often temporary and dependent on a good rainy year. The dried-up lakes are cultivated by market gardeners using a specific farming system. The main crops are cabbage and tomatoes. Other include aubergines, chillies, cucumbers, courgettes, etc. On these sites, production requires little or no watering: water is supplied by the aquifers. The layout of the plots follows the banks and stretches towards the centre of the valley or lake. Farm size rarely exceeds one hectare. On the Langue de Barbarie, crops are mainly recession crops: cabbage and sweet potatoes.

6. The diversity of vegetable production in the *Niayes*

Agriculture in the *Niayes* is largely dominated by market gardening. Food crops, cereals and other crops are very little developed. Production volumes have risen sharply over the past 15 years, to more than 600,000 tonnes. Between 2011 and 2015, market garden production in the *Niayes* increased from around 640,000 tonnes in 2011 (PADEN, 2013) to around 886,930 tonnes in 2015 (DAPSA, 2016). This is favoured by industrial farms, with the predominance of orchards and home gardens and private farms. This situation may be seen as positive for farmers, but it has serious consequences for soil resources. Thus, strong land pressure and the disappearance of fallow land are partly responsible for land degradation in the *Niayes*. This research focuses traditional market garden farms, which are in a different situation from the national average, with a decline in production as a result of alkalinisation of the soil and salinisation of irrigation water. The peri-urban system family market-garden production is the main production system, with less than one hectare of land, relatively insecure land tenure (sharecropping, renting) and a high presence immigrants from neighbouring countries.

Over the last ten years, the market garden sector has benefited from programmes and projects that have mobilised more than 10 billion CFA francs. The results achieved are convincing, but need to be consolidated and continued. The enormous efforts made are beginning to dissipate and compromise the sub-sector, due to water supply difficulties and strong land speculation on agricultural land for the benefit of property developers. In 2011, around 41% of Senegal's vegetable production came from the *Niayes* (ANSD, 2012). Many foreigners, especially Guineans, work in the market gardens. People use the income from market gardening to build permanent houses, hence its importance in the area. In the regions of Dakar, Thiès, Louga and Saint-Louis, onions dominated other crops (367,500 tonnes) in 2015, accounting for 41.43% of all vegetables (DAPSA, 2016). It was followed by tomatoes (28,000 tonnes) in 2015 and 2016, compared with 52,895 tonnes in 2018 (Dieng, 2018), which showed a marked decline compared with 2010 (over 55,000 tonnes). Vegetable production is generally good, with almost all the vegetables used by the local population being grown: potato (*Solanum tuberosum*), green bean, okra (*Abelmoschus esculentus*), chilli, aubergine, turnip (*Brassica napobrassica*), cabbage *(cabus* pommé), jaxatu (bitter aubergine), cassava, carrot and sweet potato (*Ipomoea batatas*).

Figure 8: Prevailing speculation in *the Niayes* area in 2015

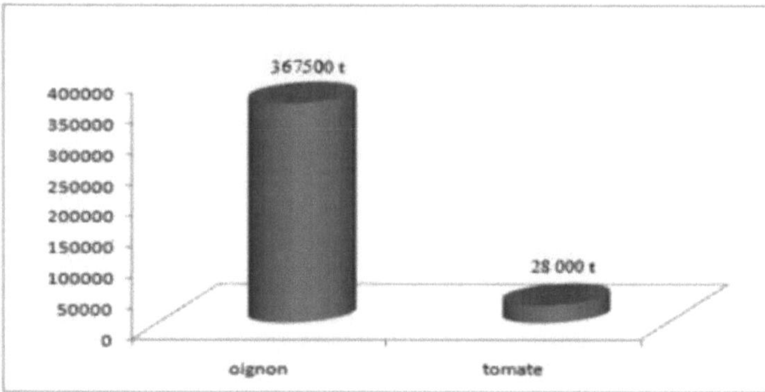

Source : Direction de l'Horticulture 2016

There was an increase in production, depending on the crop variety, between 2011 and 2013 (from over 56% for cassava to over 209% for okra). This shows the high level of speculation in market gardening, and the pressure on land, as in the Dakar region. This situation contributes to soil . Farmers are often forced use more and more chemicals (NPK) to increase their yields. Heavy urbanisation in the region is encroaching on farmland and market gardens, forcing farmers to overuse chemicals and stop leaving land fallow. However, excessive use of chemicals contributes to alkalinisation and salinisation. Figure 9 shows market garden production in the region between 2011 and 2013.

Figure 9: Market garden production in the *Niayes* from 2011 to 2013

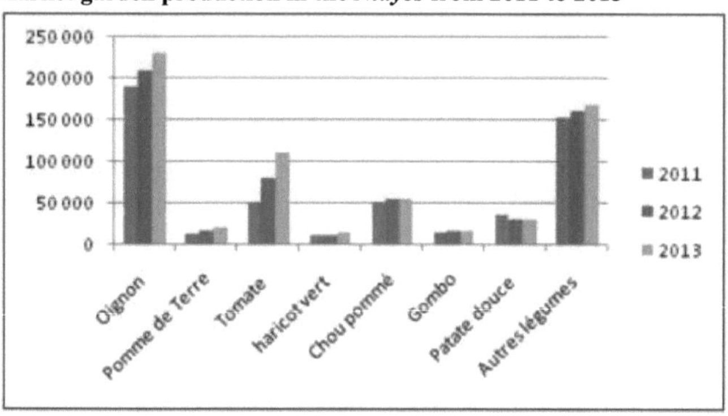

Source: Horticulture Department 2014

Conditions for the development of market gardening are favourable in the *Niayes*. Between 2009 and 2015, market garden production increased, boosted by industrial farms. Crops are diversified and intended to supply local markets and for export (for certain products). Vegetable production in 2015 is estimated at around 886,930 tonnes, an increase of 19.54% on 2014 and 25.10% on the average for the previous three years (DAPSA, 2016).

Figure 10: Market garden production in the *Niayes* in tonnes from 2009 to 2015

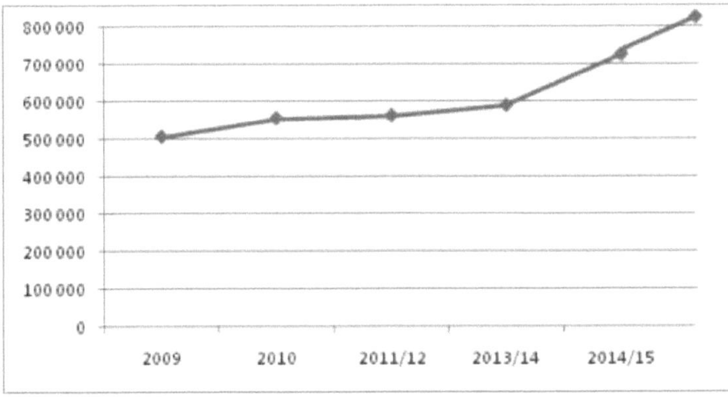

Source: CDH (2014) and DAPSA (2015)

Market garden production is highly diversified *in the Niayes* area. There are several types of crops and vegetable farms. Produce is sold all over the country and internationally, depending on market needs.

6.1. The different types of crops grown in the *Niayes*

The presence of the shallow to sub-surface water table means that there are two types of crop: flood recession crops in the *Niayes* of Dakar and Mboro and irrigated crops in the more remote parts of the depressions. In the *Niayes* of Dakar, particularly the Grande *Niayes* of Pikine and the *Niayes* of Mboro, the main flood-recession crops are tomatoes, sweet potatoes and onions, especially in Mboro. Crops requiring irrigation are dominated by cabbage and onions throughout the *Niayes*, from Dakar to Saint-Louis. The production system in the depressions is continuous throughout the year, with the cropping calendar determined more by market needs than by weather conditions.

6.2. The different types of vegetable farms recorded in *the Niayes*

There are three types of market gardening farms, depending on their size and the way they are developed:

- **Small farms,** which are less than 1 ha in size, are more of an individual than a family operation. This individual character is closely linked to the way in which the land is appropriated and developed: these are often parcels of land divided up by their owners (Lébous from the Dakar region in general), who rent them out to national migrants or migrants from the sub-region, particularly Guinea Conakry, or who practise sharecropping (locally known as "mbay sêedo"), Fall et *al.* (2001). This type of farming is dominant throughout the *Niayes*, particularly in low-lying areas and dry valleys. Characterised by their small size, very rich soils and low water stress, particularly in the Dakar and Thiès regions, the dominant crops are cabbage (*Brassica oleracea*), onion (*Allium cepa*), chilli pepper (*Capsicum*), aubergine (*Solanum melongena*), jaxatu (*Solanum aethiopicum*), lettuce (*Lactuca sativa*) and mint *(Mentha spicata*). Mint makes a substantial financial contribution, enabling the financing various inputs and meeting the daily needs of smallholders. Crops are varied, depending on the type of soil, the region and the size of the farms, and are mainly intended to supply local markets. Unlike other regions, production in Dakar is more diversified and spread throughout the year.
- **Medium-sized farms** of between 1 and 20 hectares are located on dior and vertisol soils

in the Sébikotane and Pout areas. These farms are privately owned, the owners having acquired them by inheritance, purchase, gift or lease. In terms of the way they are developed, these farms are of the modern and semi-modern type. They mechanical tools for working the land, drainage and irrigation. They employ farm workers or hired labour. Crops are less diversified than on small farms, as production is mainly intended for the wholesale market and for export (Fall et *al.*, 2001). The main crops onions, tomatoes, green beans (*Phaseolus vulgaris*), cabbage, sweet and bitter aubergines (*jaxatu*), chillies, peppers (*Capsicum annuum*), okra (grown more in the rainy season), etc.

- **Modern farms are** larger than 50 hectares and have the necessary technical and human resources. They are either associative (GIE) or private and are concentrated the regions Dakar (Sébikotane), Thiès (Pout, Mboro) and Saint-Louis. Most of their production consists of export crops (green beans, tomatoes); surpluses are used to supply local markets (Fall et *al*, 2001).

6.3. Yield trends in the *Niayes*

On a national scale, research has shown an increase in yields depending on the year since 2009, with an increase of over 19% between 2014 and 2015. This is due to intensive farming for the export of market produce. Due to alkalinisation and salinisation, yields have been falling steadily over the years in our study areas. It is in the Thiès and Saint-Louis regions that we find the largest number of farmers practising market gardening against the cold season. The plots cultivated are fairly small, varying from 1,000 m^2 to 1 ha (Fall, 2001). Plots are becoming denser and there is a greater diversity of crops. The dominant crops are still cabbage, onions, tomatoes, sweet and bitter aubergines (jaxatu in Wolof), radishes, lettuce, mint, peppers, parsley, leeks, cucumbers, etc. Onions predominate, with very high yields, varying between 150,000 and over 230,000 tonnes in 2009 and 2014, followed by cherry tomatoes, between 50,000 and 100,000 tonnes, and cabbage, between 40,000 and 100,000 tonnes. The presentation of the sites will give us a more detailed overview of the Niayes region.

7. Presentation of the study sites

The Niayes area, which runs along the northern coastline, crosses four regions: Dakar, Thiès, Louga and

Saint-Louis. What they have in common is the market gardening practised there and the problem of salinisation, which affects the sites of Lake Wouye, Lake Tanma, Taré and Darou Mboumbaye.

Figure 11: Location of the various sites studied

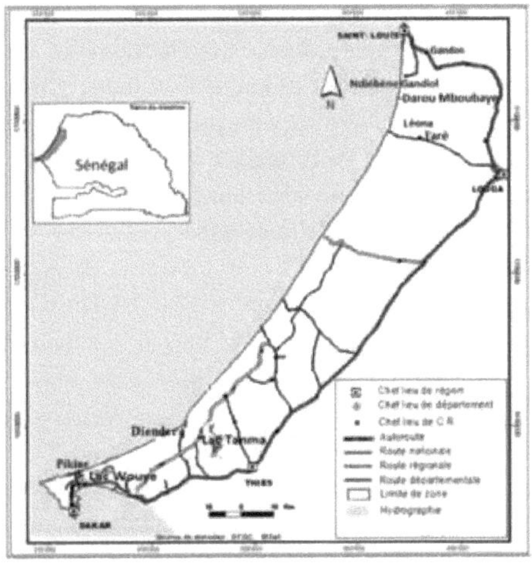

7.1. Lake Wouye

Lake Wouye is located in the Dakar region, in the department of Pikine. Covering an area of around 23.2 ha (PLD, 2006), Lake Wouye is the result of natural circulation between several watercourses that originated in Lake Thiaroye, passing through Lake Reubeut in the district of Yeumbeul, Lake Thiouroure and Lake Warouwaye in Yeumbeul. Its waters flowed as far as Lake Mbeubeuss in Malika, passing through the Toll Mbaye Djinn lagoon (Fall, 2006). These lakes, which used to communicate, are now separated by houses and roads; Lake Wouye is isolated. Most of these lakes dried up during the drought of the 1970s (Mbeubeussé, Reubeut, Thiouroure). The northern part of Lake Wouye is bordered by a filao forest, stabilising the sand dunes and preventing the site from silting up. Market gardening used to be carried out in the south-east and west of the lake. Since 2008, a large part of the market garden has been swallowed up by the water, with its transformation into a retention basin for wastewater from the surrounding neighbourhoods. In the southern part, there is a heap of household waste next to the houses.

7.2. Lake Tanma

Lake Tanma is located in the Thiès region, at (14°53'61" N) and (17°02'70" W) in the commune of Diender, arrondissement of Keur Moussa. It lies between the communes of Diender and Mont Rolland, and borders the entire north-western part of the commune to the south. The lake is also known as Lake Tamar, Lake Tamna and Lake Tanma. It is a shallow body of water of the coast, located 45 km to the east, north-east of Dakar. It stretches for around fifteen kilometres along an approximate north-south axis, with an average width of around 2.5 km, except for 113 of its length from the northern end, where it narrows to 0.750 km by 1.5 km. This constriction is accompanied by a change in the direction of elongation, which becomes north-east to south-west (Sow, 2001). "It is a brackish-water lake that partly communicates with the sea during the winter rains. The *Niayes* to the north-east of the lake are small islands" (Morel and Touré, 1967: 573). "The lake has now dried up and the paleochannel has been completely buried by the advance of the revived yellow dunes" (Sy, 2000: 10).

7.3. The Taré sector

Taré is in the Louga region, commune of Léona, arrondissement of Sakal. The village is located after Potou, on the southern edge of the ancient mouth of the Senegal River, which runs along the Atlantic Ocean for around 30 km. It is at the southern end of the Langue de Barbarie National Park (PNLB), between the villages of Gabar and Mourel. It runs along the north-western edge of the commune, with market gardening taking place in the southern, south-eastern and western parts. There are white dunes planted with filaos and coconut palms, stretching across a wide beach swept by the trade winds.

7.4. Darou Mboumbaye

Darou Mboumbaye is in the Saint-Louis region. It lies 163 km north-east of Dakar, in the commune of Ndiébène Gandiol, arrondissement of Rao. It is in latitude (15° 50' 16" North), and longitude (16° 31' 0" West) (1fovil, 2016). The study of the various sites makes it possible to determine the continuity of salinity, which remains a recurring problem in the *Niayes*.

Research objectives

The aim of this study is to determine the chemical facies of the soil and water in order to show the origin of the salinity, to compare them and to monitor their evolution over time. The scientific results obtained in this research show the quality of the groundwater in terms of its potability and suitability for market gardening. The link between salinisation and the consequences for market gardening, arable land and yields. The theoretical foundations and conceptual framework shed further light on the research problem.

Conclusion

Chapter 1 describes the physical setting of the *Niayes*, the situation of market gardening in the region and the factors contributing to the salinisation of the *Niayes*.

In order to better circumscribe the formulation of the problem, chapter 2 deals with the bibliography and the theoretical and conceptual foundations, for a better understanding of the concepts of the study.

Chapter 2

State of the art, theoretical and conceptual foundations

Chapter 2 presents the state of the art, the research problem, the research questions, the delimitation of the field of investigation, the research objectives, the justification for the choice of the theme, the interest of the research and the research hypotheses. The conceptual foundation or operating framework, refers to the definition of the concepts of the thesis: salinisation, consequence, market gardening, for a better understanding of the theme.

1. State of the art

A great deal of research has been carried out on salinisation, market gardening, the *Niayes*, and so on. These include Le Brusq and Loyer (1982), Schiess (1995), Diouf, Gomis, and Tchani (1997), Boivin, Favre, and Maeght (1998), Ndour (2001), Jacoutot and Diallo (2006), Diallo, Ndiaye, Saleh, Tine, Diop, and Guissé (2015), Niang (2017). These authors focused mainly on physiological aspects and provided details on all or part of the region. Some have focused on socio-economic activities (market gardening). However, on the whole, the research carried out in this field has produced some very interesting results that have enabled us to appreciate what has already been done.

Le Brusq and Loyer (1982), *Évolution de la salinité des sols et des eaux en relation avec la riziculture submergée dans le delta du fleuve Sénégal*, ORSTOM, 16 p, determined the influence of soil and microbiological parameters on productivity in the Senegal River Delta. They distributed the salt-affected soils on the various 1:200,000 maps of the valley (FAO, 1973). They then studied changes in soil salinity and pH using an experimental protocol in which they chose 4 plots representative of salinity levels, soil types and geomorphological positions in the irrigated basins. In relation to our theme, the various analyses carried out in this research were very interesting for the study of salinity: pH, electrical conductivity (EC). Attempts were made to assess the water and salt balance in the Tilene, Pont Gendarme, Ngomene and Ndiaye basins. Better management of the consequences would have been a plus.

Schiess (1995), *Alcalinisation des sols irrigués dans la Moyenne vallée du fleuve Sénégal: diagnostique par évaluation du stock de carbonates,* Diplôme EPFL : Génie Rural : Pédol, Lausanne, ORSTOM, 39 p., tested a simple method for assessing the stock of carbonates accumulated in irrigated soils in the Moyenne vallée du fleuve Sénégal. This work enabled him to show that there is an evolution in cultivated soils, manifested by a rise in pH and an accumulation of carbonates in cultivated soils. The study shows that all irrigated soils undergo the same changes, to a greater or lesser extent depending on the age of the irrigation scheme, the water regime and the type of soil. No area has been spared, and even areas equipped with drains are deteriorating. The study showed that the soil is becoming more alkaline. Schiess points out that the phenomenon of alkalinisation is troublesome, insofar as it affects soil productivity. The results of the study showed yield reductions observed by farmers in the region, but Schiess says that these are not solely due to the phenomenon of alkalinisation, as many other factors may be to blame. The fact remains, however, that the accumulation of carbonates causing the pH to rise, along with high alkalinity, will affect soil fertility over the long term. Alkalinity causes ionic deficiencies and toxicities that can disrupt the plant growth cycle. The results showed that all cultivated soils follow the alkalinisation pathway to varying degrees. While these results have given us an idea of the alkalinisation of soils and the drop in yields, through a reduction in soil fertility, we still need to find a simple and effective means of combating the problem, compatible with the cultivation techniques and financial resources of Senegalese farmers.

Diouf, Gomis et Tchani (1997), *Le contact eau douce-eau de mer entre Dakar et Saint-Louis*, colloque, ORSTOM, Laboratoire de Géophysique, 138 p, show the hydrogeological interest of the electrical method which could be used to follow evolution of the equilibrium of the coastal aquifers in the *Niayes* area. This is an interesting geophysical study using electrical soundings, carried out along the coast between Dakar and Saint-Louis, as part of the "Gestion des nappes des *Niayes*" project. The study determined the morphology of the interface between freshwater and saltwater aquifers along the coast in order identify areas vulnerable to invasion by seawater. The aim is to manage the aquifer by constantly monitoring salinity. In this research, the authors did not highlight the consequences for market gardening activities in the area, which would have been interesting.

Boivin, Favre, and Maeght (1998), *Les sols de la Moyenne vallée du fleuve Sénégal : caractéristiques et évolution sous irrigation,* article, ORSTOM, Institut Français de Recherche Scientifique pour le développement en Coopération, École Polytechnique Fédérale de Lausanne, IATE-Pédologie, Lausanne, 245 p, attempt to answer the question of the impact of irrigated crops on the soil resource. They partially reconstructed the evolution of soils under cultivation by comparing bare soil and cultivated soil. The study clarified the water balances and geochemical changes in solutions through detailed functional analyses. The results suggest that the soils are partly on an alkaline, saline path. This study provides additional information on the relationship between irrigation and soil degradation. The consequences of salinity on agricultural activities, such as market gardening, also remain to be demonstrated.

Ndour (2001), *La dégradation des sols au Sénégal : l'exemple de deux communautés rurales (Kaymor et de Mont Rolland)*, dissertation, UCAD, 314 p, takes stock of the determinants of soil degradation in Senegal, over a period from 1988 to 2000, using demographic and agricultural data and Landsat TM image analysis. The study shows that agricultural yields remain extremely low. The study's consideration of the impact of soil degradation on agriculture was very interesting. However, the report did not quantify the level of soil degradation using scientific data.

Jacoutot (2006), *Modifications environnementales et conséquences dans le Bas-delta du fleuve Sénégal,* dissertation, Université Joseph Fourier, 174 p, discusses the socio-economic activities that depend to a large extent on natural resources in the deltaic environment. The study relates socio-economic activities such as market gardening to environmental changes. It gives good results on the salinisation of land and water in the Gandiolais. It's a highly relevant study in terms of scientific findings on the salinisation of soil, surface water and groundwater, with piezometric data showing a drop in the water table in the quartz sands in the northern *Niayes* area. Data on the decline in agricultural yields do not appear in this report.

Diallo (2006), *Aménagements hydrauliques et mutations du Bas-estuaire du fleuve Sénégal : Impacts sur la nappe des sables dunaires du Gandiolais,* dissertation, UCAD, 79 p., discusses the main impacts of hydraulic developments on the surface water table in the Gandiolais. The results of the study show a drop in piezometric levels and the salinity of the water table, as well as the combined effect of unfavourable climatic conditions. The water analyses carried out show that the salinity comes mainly from irrigation water. The results confirmed by our study could be reinforced by the consequences of salinisation on agricultural activities, particularly market gardening.

Diallo, Ndiaye, Saleh, Tine, Diop, and Guissé (2015), in *Étude comparative de la salinité de l'eau et des sols dans la zone Nord des Niayes (Sénégal),* Science Journal, Vol. 23, pp. 101-111, assessed the acidity and salinity of irrigation water and soil, with seven soil profiles that were described and their granulometry determined. The study showed that salinity is higher at the surface than at depth, in the northern zone of the Niayes, with two sub-zones: a continental part subject to higher acidity and a coastal part subject to high salinisation, which considerably limits agricultural production, but data on the consequences for farming activities would have been interesting.

Niang (2017), *Dégradation chimique et mécanique des terres agricoles du Gandiolais (littoral Nord du Sénégal), analyse des dynamiques actuelles d'adaptation*, thesis, UGB, 355 p., showed through *in situ* measurements the EC and water quality, based on the hydrochemical approach. Based on hydrochemical data, he was able to highlight high groundwater salinities in the coastal fringe, with some anomalies inland. The results of the analyses show that the salinisation is due to sulphates, chlorides, calcium magnesium, sodium and potassium. The study shows that the origin of these elements is directly related to past and present intrusion, and that irrigation with these waters should be avoided for a large part of the Gandiolais region. The geochemical approach made it possible to determine the low salinity of the exploited soils, which do not retain the various salts observed in irrigation water. Soil analyses (in dry and rainy seasons) did not show any concentration of salts in this sandy complex. However, moderate salinities were noted at certain measuring points, indicating a certain subsurface concentration. The study shows that the salinity of the soil solution is the main mechanism for increasing the osmosis effect, which leads to very low vegetable production. Hence the relevance of this study, which uses scientific data to show that the salinity of the Niayes comes mainly from the water and not the soil. It shows that the deterioration in market garden production conditions is the sum a multitude of forces, of which the deterioration in water quality is the main factor in the current sluggishness of the market garden system in Gandhi. A comparison of the results obtained gives an idea of how the situation is evolving and its consequences for market gardening activities.

The major contribution of these documents lies in their ability to draw attention to the salinisation of soil and water, which is increasing on a national scale. These authors have tried to provide solutions by posing the problem. Through these studies, we learned about the realities on the ground. In any case, the consequences of salinisation on market gardening have not provided satisfactory statistical data, within the framework of the state of the art. The theoretical foundations shed more light on the position of this study, which, through the scientific data obtained, shows the consequences of salinisation on market gardening activities.

2. Theoretical foundations

2.1. Issues

The *Niayes* provide around 60% of Senegal's vegetable production (SONED, 2013). In recent years, market gardening in the area has been compromised by rising salinity, exacerbated by pressure on the water table. Peri-urban agriculture has become an essential, high value-added activity in supplying fresh produce to the major cities. In the *Niayes* region, this type of agriculture, developed in a context of high water deficit and soil poverty, is practised intensively as a result of strong demographic growth and rapid urbanisation. Excessive use of chemicals and pressure on the land (with the disappearance of fallow periods) are leading to falling yields and the abandonment of several plots.

In 2018, the *Niayes* are still suffering from a problem of salinisation, with a reduction in market gardening areas, a situation felt in our sites: Lake Wouye, Lake Tanma, Taré and

Darou Mboumbaye. In these areas, market gardens are disappearing, leading to a reduction in arable land. The market garden areas are affected by salinity, leading to the abandonment of several plots. On those plots that are still being farmed, production has fallen in recent years. In Gandiolais, the proximity of the Langue de Barbarie (LB), which was an asset for the development of market gardening (shallow water table), has become a handicap since the opening of the breach in the LB in 2003. The breach caused salt to intrude into the water table, polluting the wells used for market gardening. The salinisation of the water table is having an impact on vegetable production, with the alkalinisation of the soil making it less and less fertile for market gardening. The rise in the freshwater table has given way to increasingly salty water, a threat to market gardening. In recent years, the drop in rainfall has had a negative impact on the water table, which has fallen (causing fossil salts to rise by thermocapillary action), limiting market gardening in the area. The worsening climate in recent years, combined with over-exploitation of the NSQ (to the south), has led to a drop in the water table in some areas.

A large number of market garden plots are being abandoned due to infertility in 2018. In our study areas, many market gardeners have preferred to abandon their plots, which are no longer profitable. Climate change has caused the sea to advance towards the coast, with the intrusion of the salt wedge into the NSQ. Market garden plots located near the sea and the Senegal River are increasingly subject to saline intrusion, which has repercussions on market gardening activities and therefore on yields.

This research, carried out in the *Niayes* region (Dakar, Thiès, Louga, Saint-Louis), focuses on salinisation and its effects on market gardening.

2.2. Research questions

The main research question is: What are the saline dynamics of the Niayes and their consequences for market gardening? The main question is subdivided into two secondary questions:

- What causes salinisation?
- What are the consequences of salinisation on market gardening activities and yields? We will use the research questions to define the scope of the research.

2.3. Defining the field of investigation

This research is limited to the Great Coast of Senegal, which covers the area of the Niayes, from the Dakar region to the Saint-Louis region. This makes it possible to understand the dynamics of salinisation in the *Niayes* and its impact on market gardening. Since the 1970s, the *Niayes* have had to contend with environmental problems (drought, falling rainfall, falling water table, alkalinisation of the soil, salinisation of irrigation water, etc.), which have had an impact on market gardening (the main activity in the area). The objectives of the research were defined by defining the field of investigation.

2.4. Research objectives

The main aim of this thesis is to study salinisation in the *Niayes* and its impact on market gardening .

Under the main objective, we have set the following specific objectives:

- study the causes of salinisation in the *Niayes* ;
- analyse the consequences of salinisation on market gardening.

Justifying the choice of theme showed the interest of this research.

2.5. Justification for the choice of topic and research interest

Since the 1980s, much interest has been shown in salinisation in the *Niayes* zone (Diakhaté,

1986). Studies have been carried out in this context by (Fall et *al*, 2001; Diallo, 2006; Ba, 2008; Diallo et *al*, 2009; Ndiaye et *al*, 2012; Diallo, Ndiaye, Saleh, Tine, Diop and Guissé, 2015; Niang, 2017; Sène, Sarr, Kane and Diallo, 2018), providing scientific results on salinisation or alkalinisation, through in-depth analyses of natural resources (soil and water). The choice of theme is favoured by the results of our DEA dissertation, which showed a degradation of agricultural land at Lake Wouye in Malika in 2007. This aroused interest in the degradation of farmland in the *Niayes*. This study identifies the problem of salinity, which threatens the survival of market gardening activities in the *Niayes* across the four regional sub-groups (the *Niayes* of Dakar, Thiès, Louga and Saint-Louis).

Our work seeks to understand the evolution of the phenomenon of salinisation in the *Niayes*, with a view to a larger synthesis. In addition, the *Niayes* are an area where many issues are at stake. They play an important role in the national economy because of their diverse potential: market gardening, poultry farming, etc. However, market gardening remains the main activity in the traditional production system. The current study of the economic consequences of salinisation is helping us to understand the constraints on market gardening areas, and to identify the key factors that need to be addressed to preserve the *Niayes* ecosystem, which is the main source of survival thousands of farmers. Our study sites are important areas from an environmental, social and economic point of view, hence the importance we attach to them. The development of market gardening has led to the alkalinisation or salinisation of soil and water. The *Niayes* were chosen for a number of reasons, the most important of which are :

- the over-exploitation of natural resources such as water and soil;
- the excessive use of chemicals in agriculture ;
- the advancing sea due to climate change; - salinisation.

All these factors justify our interest in this environment and in the development of market gardening, hence the formulation of the research hypotheses.

2.6. Research hypotheses

The main hypothesis is as follows: the process of salinisation of the Niayes has negative consequences on market gardening activities. Two secondary hypotheses follow from the main hypothesis:

- Salinisation of the *Niayes* results from both natural and man-made causes.
- the salinisation of the Niayes has led to a significant drop in market garden yields.

To better understand the theme of this research, the key concepts are defined.

3. Conceptual foundation (operating framework)

3.1. Conceptual definition

For a better understanding of the subject of this research, certain concepts have been elucidated: salinisation, consequences, market gardening.

Salinisation is an increase in the salt content of soil or water that is likely to have a negative impact on vegetation or agricultural activities (market gardening). It is the enrichment of a soil in soluble salts, leading to the formation of a saline soil (Hosni, 2009). The phenomenon is more active in soils with little or no leaching and intense evaporation (arid regions), or in the low-lying parts of coastal areas in soils with low permeability (dictionnaire de l'agriculture, 2013). This is an accumulation of salts in the soil, which reduces yields and can irreparably damage agricultural land (FAO, 2002). Salinisation of a soil involves excessive enrichment in soluble salts (chlorides, sulphates, carbonates, etc.) of sodium, magnesium or calcium (König, 2006). Soil is considered salty when its soluble salt content exceeds 1 to 2%

in the top 20 centimetres (Mainguet, 1995). This is the process by which salts accumulate in the soil. Soil salinity affects plant growth (vegetative cycles) by reducing their capacity to absorb water (Eilers, al, and Sampson, 1995).

In our case, salinisation can be considered as the concentration of soluble salts in the soil, which in large quantities has negative consequences on agricultural activities, justifying a drop in yields. This is why it is so important to understand the concept of consequence, in order to better understand the effects of salinisation on agricultural land.

Consequence is the result of a set of events brought about by an action. It is necessarily produced by something, the logical result of a cause or an action. In the case of this research, it is the effects induced by soil salinisation, which are likely to partially or definitively destroy market gardening areas in the *Niayes* zone, with the loss of several hectares of arable land and a drop in yields. Understanding the concept of market gardening will provide a clearer picture of the consequences of salinisation on market gardening activities, justifying the definition of the concept.

Maraìchage is the growing of vegetables, fruit, herbs and flowers for food, for personal or professional use. It can be intensive or traditional, outdoors or under cover. Market gardening in the *Niayes* area increased with the drought of the 1970s, which prompted farmers to consider switching to market gardening (Enda Graf, 2003). Market gardening in *the Niayes* area is a way of making the most of the land located in the basins, with small farms. It plays an important role in people's activities. It is practised as a means of supplementing income, combined with other activities or as the main activity. It is a potentially profitable crop, thanks to a dynamic national market. In our study areas, it is practised in the traditional way, using rudimentary hand tools. The variables and indicators will shed more light on the operationalisation of the concepts.

3.2. Putting concepts into practice

The operationalisation of concepts involves choosing variables and indicators before showing the relationship between hypotheses and variables.

Table 1: Choice of variables and indicators

Concepts	Dimensions	Variables	Indicators
Causes of salinisation in the Niayes	Anthropogenic	Use of chemicals, pumping of the water table, widening of the breach south of Saint-Louis	Quantity of chemicals used, piezometric level, width of breach (Spatial spread of marine intrusion)
	Natural	Drought in the 1970s, variability in rainfall, significant evapotranspiration, active coastal erosion.	Quantities of rainfall per year, evaporative demand, sea level.
Consequences of the salinisation of Niayes	Social	Soil productivity, arable land, agricultural yields, land abandonment, pumping from wells far from degraded land, addition of organic matter (OM)	Areas lost or abandoned (in ha), Cultivable area, production in tonnes per area, volume of water/day, quantity of OM

The analysis model sheds more light on our study.

4. Analysis model

The conceptual analysis model on salinisation and its consequences is integrated into the methodological approach.

Figure 12: Conceptual analysis model

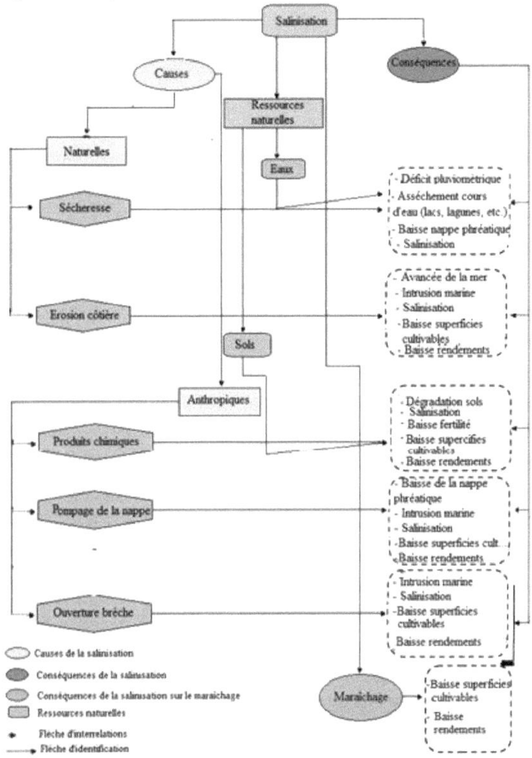

The analysis model brings together the causes of salinisation and the consequences, with the long-term drop in yields. The model identifies the resources affected (water and soil) and the vulnerable economic activities (market gardening), so that the assumptions can be better integrated.

Conclusion

Chapter 2 (State of the art, theoretical foundations and operating framework) sets out the research problem, objectives and hypotheses, in order to move the study forward. Chapter 3 deals with the methodological approach and its various stages.

Chapter 3

Methodological approach (tools and methods)

The methodological approach concerns tools and methods. It focuses on field data collection, with soil and water sampling for laboratory analysis and data processing. It plays an important role in this research, focusing on a detailed study of the various actions carried out in the course of this work. It is based on a combination of qualitative and quantitative methods, the first stage of which is the collection of field data.

1. Field data

In the field, several stages were carried out at different times, in 2010 and 2018. The first stage was carried out in the *Niayes* of Pikine in the Dakar region (Commune of Malika) in April 2010 to observe the evolution of the situation in relation to the research already carried out in 2006 and 2007 as part of our master's and DEA thesis. The first study site was Lake Wouye in Malika. A preliminary survey was carried out to determine whether the salinity problem observed in the *Niayes* of Pikine in Dakar was general or localised. For this reason, after our DEA research (2007), the site was revisited in 2010 to carry out our first surveys of other regions, with the aim of obtaining information on salinity problems and selecting study sites.

This initial fieldwork provided theoretical evidence that the other regions of the *Niayes* were also facing salinity problems. Theoretical information needed to be verified scientifically, which is why soil and water samples were taken. This work does not differ from studies already carried out in the *Niayes* area, and our data is compared with scientific data obtained at earlier dates, as well as more recent ones. It appears that the environment studied is very dynamic, as is the phenomenon studied. For this reason, we have compared our results with those already obtained, in order to assess the current dynamics. Following the same logic, we carried out a second fieldwork activity in July 2018, to carry out further soil and water sampling in order to have two reference years in which to compare the results. These field activities in 2010 and 2018 provided quantitative data (soil and water samples, interviews) and qualitative data (laboratory analysis results).

2. S es de prélèvements

The sampling sites were chosen in the vicinity of Lake Wouye, Lake Tanma, in the market gardens of the village of Taré in the commune of Léona and in the village of Darou Mboumbaye in the commune of Ndiébène Gandiol. For all four sites, we worked with two 15 cm horizons per benchmark. For Lake Wouye, samples were taken in 2007, 2010 and 2018, starting with the plot closest to the lake.

In Lake Tanma, samples are taken from the fields towards the lake. In the village of Taré, located next to the Langue de Barbarie, the reference point is the position of the river in relation to the plots. In the village of Darou Mboumbaye, samples are taken from plots to the south of the village, to the east. In Darou Mboumbaye, Dické D. (2018) reports that the plots sampled in 2010 and 2018 (including his own) are located around 2 km from the Senegal River and 2.5 km from Lake Thième, where salt is collected during the dry season. It is important to know the distance between the plots sampled and the river, in relation to the effects of the breach in the Gandiolais.

Photo 3: Soil sampling in 2010 and 2018

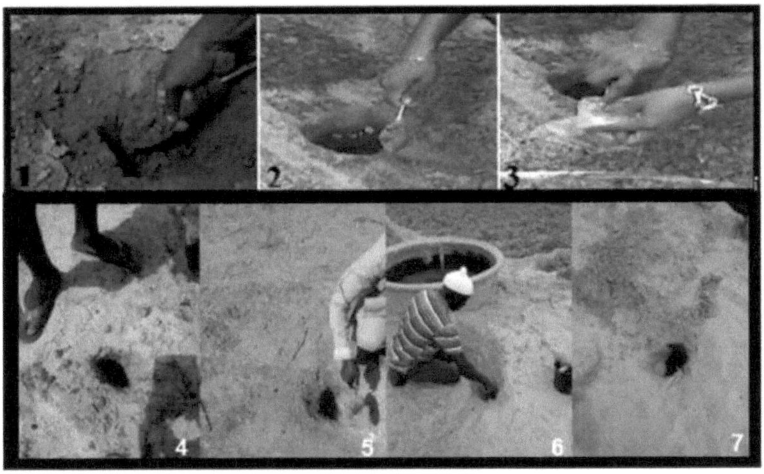

Source: Photos by M. Fall (June 2010 and July 2018)

The photos show samples taken in 2010 and 2018. Images 1, 2 and 3 show samples taken in 2010 (Lake Tanma, Taré and Lake Wouye) and images 4, 5, 6 and 7 show samples taken in 2018 (Taré, Lake Tanma, Darou Mboumbaye and Lake Wouye). The same sampling techniques were used at all four sites.

3. Sampling

The study of salinisation required seasonal observation: dry season, rainy season. The soil and water sampling strategy was based on seasonal contrasts. Sampling of soil and water (lakes, wells, river) was carried out in the dry seasonduring April, May and June 2010, to measure pH and electrical conductivity (EC), at the SAED soil fertility laboratory.

In 2018, samples taken during the rainy season, in July, and then analysed at UCAD's hydrochemistry laboratory.

In 2010, soil and water samples taken from the study sites were analysed to determine pH and electrical conductivity (EC) values.

In 2018, a full chemical analysis of the water samples was carried out to determine the chemical facies of irrigation water. This choice was made on the basis of the 2010 results, which justified a full chemical analysis of the water.

Soil sampling[4] was carried out on cultivated and uncultivated (abandoned plots) market-garden areas due to salinisation, using a trowel and a tablespoon. In the Dakar and Thiès regions, plots were chosen according to their proximity to the Wouye and Tanma lakes. In the Louga and Saint-Louis regions, the choice was guided by the plots under cultivation. Uncultivated or abandoned plots were chosen according to their proximity to cultivated plots. For all samples, targeted sampling (judgmental type) was carried out. The type of sampling chosen takes into account the influence of lakes, the river and the saltwater wedge on the

[4] As a reminder, from 250 mho onwards we have the beginning of salinity for electrical conductivity: CE. The ohm is used to measure salinity, and its opposite is the mho, which is expressed in mmho (millimho). Mho = 1 mmho/cm / Ohm (if it's equal to 250 mho/cm), there's the beginnings of salinity. Rice can withstand up to 2000 mho/cm due to a reagent problem.

chemical facies of irrigation water. At each of the sites, 7 benchmarks were chosen, each 10 m apart in 2010, to read changes in pH and EC as the distance from the first point increases, and 3 benchmarks, at least 100 m apart in 2018. A 30 cm well was dug at the benchmarks. For surface sampling, the levels are 0 to 15 cm and 15 to 30 cm for deep sampling, corresponding to the root volume of market garden crops: onions, chillies, cabbage, carrots, aubergines, etc. This gives a total of 56 samples in 2010 and 24 samples in 2018. All the samples were of the composite type, where we chose sampling sites randomly, taking into account plots abandoned due to salinisation and those still suitable for cultivation. The water samples (6 in 2010 and 12 in 2018) were taken from wells on cultivated and uncultivated plots, on Lake Wouye and the Senegal River. Samples were taken from the Mboumbaye lagoon (river) at Taré (the river is close to the village). For the physico-chemical analyses, the samples were taken on the reverse side of the barrier beach and on the mainland in order assess the level of diffusion of the saltwater wedge.

3.1. Sampling techniques and sample units

Two types of soil samples were taken in 2010 and 2018, with 100 g/bag. In 2018, samples were taken from the same plots as in 2010. Figure 13 shows a schematic diagram of how the samples were taken.

Figure 13: Soil sampling diagram

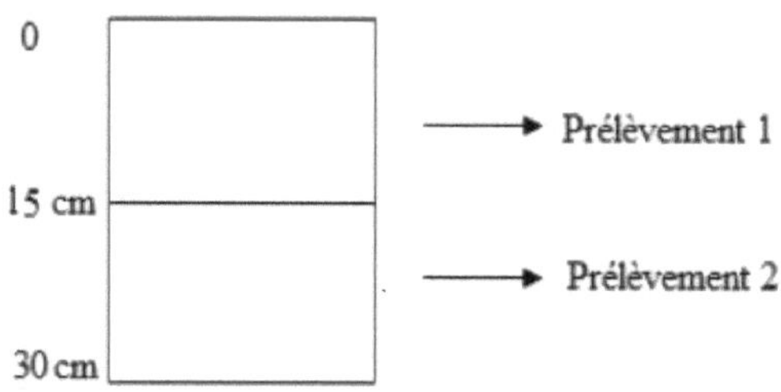

The water samples were taken carefully to minimise the risk of contamination, using clean bottles. A total of around 30 litres of water were taken from wells on cultivated and uncultivated plots, in Lake Wouye and in the Senegal River, in 2010 and 2018. In 2010, a total of 12 litres of water were taken (Lake Wouye 1.5 litres, Lake Tanma 3 litres, Taré 4.5 litres, Darou Mboumbaye 3 litres) and 18 litres in 2018 (Lake Wouye 3 litres, Lake Tanma 4.5 litres, Taré 6 litres, Darou Mboumbaye 4.5 litres). For safety reasons, water samples were not taken from Lake Tanma in 2010, due to the difficult access and unstable ground around it (too slippery, risk of sinking). In July 2018, the lake was completely dry. Our well water samples were taken from the same plots where the soil samples were taken. They were kept in clean 1.5-litre bottles and transported to the SAED laboratory in 2010, and to the UCAD hydrochemistry laboratory in 2018, with all the information concerning the sampling: site, location, date, time.

Figure 14: Sampling points

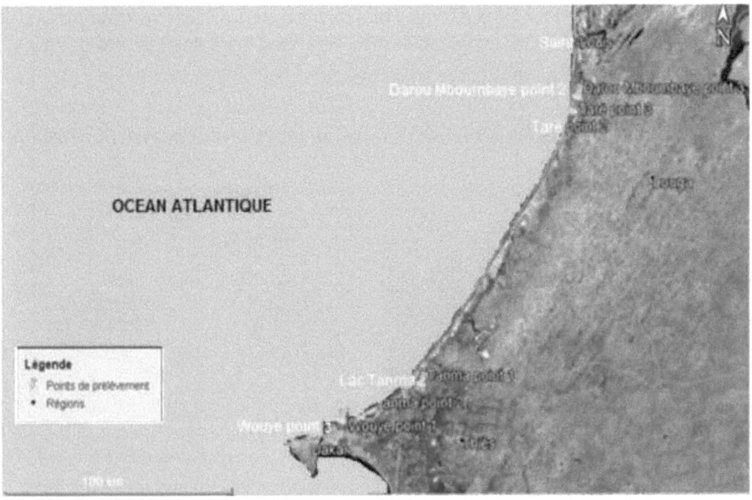

3.2. Data processing and analysis

The physico-chemical parameters measured (soil and water) were EC and pH in 2010, except for lake water, where we added bicarbonate, chlorine, sulphate, nitrate, sodium, potassium, magnesium and calcium (chemical profile of the water). In 2018, bicarbonate, chlorine, sulphate, nitrate, sodium, potassium, magnesium and calcium (chemical profile of irrigation water) are measured in the laboratory.

The quantitative and qualitative methods used in this research are based on two analytical approaches. The results of the interviews with the market gardeners were statistically and graphically processed, particularly in terms of area and yield. For surface areas, the number of plots is used to calculate them. The areas lost in 2019 are calculated on the basis of the number of parcels lost per year and the area per parcel. Secondary data from the state of the art, direct observation or interviews were analysed and processed as they came in.

3.3. Laboratory treatment

As part of the laboratory processing, a number of steps were taken to analyse the samples taken from the sites.

Photo 4: Drying of samples (conditioning)

Source: Photographs by M. Fall (September 2007 and October 2010)

To analyse the samples, the pH is determined using a sample of 20 g of sand and 50 ml of distilled water. The electrical conductivity (EC) is measured by taking a sample of 20 g of sand and 100 ml of distilled water (which will be 5 eS/cm), plus distilled water, which is left to stand for 30 minutes after mixing the water and sand thoroughly. Set the pH meter to 25°C and set to pH.

Photo 5: pH analysis at the SAED laboratory using the conductivity meter in 2007 and 2010

Source: Photographs by M. Fall September 2007, October 2010

For the EC, another 50 ml of distilled water is added to the samples and the conductivity meter is used to measure it.

Photo 6: EC analysis at the SAED laboratory, using distilled water in 2010

Source: Photo M. Fall October 2010

In 2018, the analyses were carried out at the hydrochemistry laboratory of Cheikh Anta Diop University in Dakar (UCAD).

Photo 7: Processing samples at UCAD's hydrochemistry laboratory in 2018

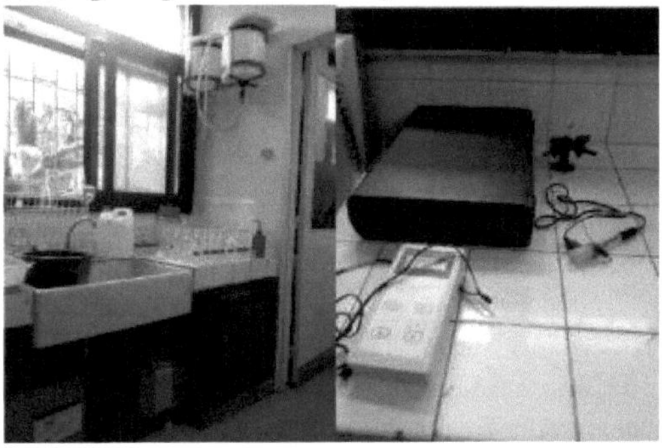

Source: Photo M. Fall July 2018

The Bocoum scale (2004, Table 2) was used to assess the pH and salinity (EC) results of the soil samples.

Table2: Assessment of pH and salinity (Bocoum, 2004)

pH		Electrical Conductivity (EC)	
pH ranges	**Soil**	**CE ^s/cm)**	**Soil**
< 4,5	Extremely acidic	< 250	Non-saline
4,6 - 5,2	Very acidic	250 - 500	Slightly salty
5,3 -5,5	Acid	500 - 1000	Salt
5,6 - 6,0	Moderately acidic	1000 - 2000	Very salty
6,1 - 6,6	Slightly acidic	>2000	Extremely salty
6,7 - 7,2	Neutral	-	-
7,3 - 7,9	Slightly alkaline	-	-
8,0 - 8,5	Alkaline	-	-
>8,6	Very alkaline	-	-

Source: Diallo M. D et *al* (2015)

To interpret the pH, we took into account the acid-base status of the project's soils UNDP/FAO (table 3).

Table 3: Acid-base status of soils according to the UNDP/FAO project

Scale	0	2	3	4	5	6	
pH	4 4.5	5	5.5		6 6.5 7 7.5 8 8.5		
Degree	Very acid	Acid		Slightly acidic	Neutr e	Low alkali n	Alkaline

Source: UNDP/FAO project, 1973

Water treatment was carried out at the SAED laboratory in 2010 and at the UCAD hydrochemistry laboratory in 2018. The pH was measured using a pH meter that had already been calibrated. The electrode is immersed in the solution to be analysed. The solution is gently stirred and left in contact for 20 to 30 seconds. The operation is repeated up to three times taking the reading. Wait until the pH is stable (between 30 seconds and one minute). For electrical conductivity, the conductivity meter should be calibrated[5] with a KCL solution

4Calibrate: Chcck (a measurement) by comparison with a standard (buffer solution which maintains

of known concentration and known conductivity. The electrode is immersed in the sample, so that the conductivity of the sample can be read directly from the instrument in ^S/cm or (mS/cm). After each measurement, the electrode is rinsed. Readings are taken at a constant temperature of 25°C. Statistical and graphical processing will shed more light on the laboratory results.

3.4. Statistical and graphical processing

Statistics and graphs were processed using Excel spreadsheets and diagrams software. This made it possible to produce diagrams and other graphs, followed by comments to support the analyses. Data from laboratory analyses are presented in Schoiller Berkaloff and Piper diagrams, to show the chemical facies. Data from *in situ* surveys, such as areas lost and yields, are presented in strip diagrams, to make them easier to read. For the mapping, we worked with geographic information systems (GIS). Some of the maps were produced by the DTGC in Dakar.

3.5. Field surveys

The field surveys provided more information on soil salinisation at the target sites. They made it possible to find out whether market gardeners are confronted with salinisation and what the consequences are for market gardening. To carry out our surveys properly, the data collection strategy took place in the communes of Malika and Diender, and in the villages of Taré and Darou Mboumbaye, with market gardeners. This part was carried out by administering an interview guide for the NGOs (Enda graf, Millennium village) and the market gardeners, to find out about the salinisation situation in the *Niayes* and its consequences for market gardening.

Interviews were organised in the regions of Dakar, Thiès, Louga and Saint-Louis using an interview guide, with partners who support local development programmes and some market gardeners. Interviews were conducted with the various players, using interview guides sent to the market gardeners, to obtain information on market gardening activities. They focused mainly on salinisation, market gardening, yields, areas lost and prospects for combating salinisation. They were drawn up taking into account the hypotheses and indicators, in order to provide answers to our various research questions. Data was collected from market gardeners in collaboration with the existing MSEs (3) and village chiefs, through interviews. The reason for working with the MSEs and village chiefs is that most of the market gardeners are members of the group and face the same problems in relation to market gardening. The MSEs and village chiefs are resource people who know all the problems of market gardening in their area. If they are unavailable, the village chief puts us in touch with people who can answer our questions. Interviews were conducted with development agents who work in collaboration with market gardeners, such as Enda Graf, ARD and Millennium Village, to find out their opinion on the level of soil degradation and salinisation on the sites, the action already taken in this area or in progress, and the prospects and strategies for combating salinisation.

Conclusion

Chapter 3 provided the necessary information for this thesis. The literature review and data collection were very useful. The processing of the data collected through soil and water sampling and field surveys provided satisfactory results for achieving our research objectives. The methodological part was the foundation of our work. Thanks to the state of the art, we

---

approximately the same pH despite the addition of small quantities of an acid or base).

had extensive documentation for our research. This part enabled us to find out what had already been done, and what we could contribute to this scientific process, to further develop the results already obtained. In addition to the visual observations made through direct observation, the soil and water samples taken and analysed confirmed certain assumptions made during the field visits. The fact is that national agriculture, particularly in the Niayes, is threatened by salinisation. It is essential to identify the causes of this situation and determine the level degradation of market gardening land. It is important to present and describe the results obtained as part of this research, hence the importance of Chapter 4 of our study (presentation and description of results), which allows us to see the results obtained thanks to Chapter 3 (the methodological approach), with the aim of verifying our research hypotheses.

Chapter 4

Presentation and description of results

The market gardening sector is very dynamic, with production levels that indicate its viability. There are a large number of players (market gardeners, NGOs), each playing a key role in the smooth running of the sector. Each of them is an important link in the production chain. What's more, it's a sector that is performing very well economically and financially. However, the sector faces a number of difficulties, such as the salinisation and alkalinisation of land and irrigation water. The sector's real competitiveness will be clearly seen in its ability to deal with the problems of salinity and alkalinity, through a strategy of adaptation. The salinisation process is more marked by high temperatures throughout most of the year, coastal erosion, limited drainage and low rainfall, which are insu.sant for leaching soluble salts from the soil. The presentation and description of the results show the situation of market gardening and salinisation in the study areas, as well as a summary of the results.

1. Market gardening activities in the study sites

Market gardening is the main activity around Lake Wouye and in the villages of Taré and Darou Mboumbaye. Around Lake Tanma, it is the only activity practised. In our various study sites, there are several types of crop, which vary according to the season and the availability of land and water.

1.1. The different types of crops on the plots

- The case of Lake Wouye

In 2006, several crops were grown around Lake Wouye, with onion (*Allium cepa*, 18%), aubergine (*Solanum melongena*, 15%), mint (*Mentha sp*, 13%), cabbage (*Solanum aethiopicum Kamba*, 9%), bissap (*Hibiscus sabdariffa*, 8%), Chinese onion and green onion (*Allium fistulosum*, 6%) predominating.

In 2007, several crops were recorded, with onion (*Allium cepa*, 17%) dominating, followed by mint leaf or nana (*Mentha sp.*, 12%), bissap (sorrel: *Hibiscus sabdariffa*, 8%), green onion (*Allium cepa*, 6%) and cucumber (*Cucumis sativas*, 4%).

Figure 15: Speculation around Lake Wouye in 2006 and 2007

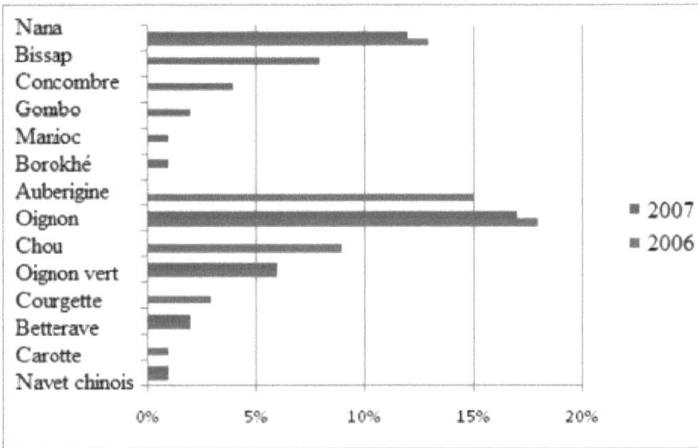

Source: Master's and DEA dissertation survey 2006 and 2007

Apart from the above-mentioned crops, there are other crops that farmers usually grow: green beans (*Phaseolus vulgaris*), sweet potatoes (*Ipomea batatas*), chillies, green chillies (*Capsicum*), leeks (*Allium porum*), peppers (*Capsicum annuum*), lettuce (*Lactuca sativa*) and tomatoes (*Lycopersicum esculentum var.*).

Photo8: Chilli pepper and aubergine plantation at Lake Wouye in 2006

Source: Photo M. Fall July 2006

Images 1 and 2 show the chilli and aubergine plantations found at Lake Wouye in 2006.

It is important to recall the situation of market gardening activities around Lake Wouye in 2006 and 2007 (master's thesis and DEA). For more regular monitoring of the first site (Lake Wouye in Malika), we returned to the field in September 2010 to assess how the situation had changed since 2007. Some 90% of the plots we worked on had disappeared under the waters of the lake, due to its transformation into a retention basin since 2008.

Photo 9: Tomato and aubergine plots destroyed by lake water and salinisation in 2010

Source: Photo M. Fall September 2010

Images 1 and 2 show tomato and aubergine plantations flooded by the waters of Lake Wouye. Soil sampling and analysis showed traces of alkalinity, with pH values ranging from 7.62 to 8.30. It should be noted that the lake water was extremely salty in 2010, with an EC of 26680 μS/cm.

The areas targeted by the 2010 study have not seen much market gardening, with many plants in poor condition. This is certainly due to the chemical properties of the soil and irrigation water. The transformation of the lake into a retention basin has had very negative consequences for market gardening in the area (flooding of plots). The situation has mainly affected lowland plots around the lake (the majority of plots). Vegetable plantations in 2010 (in the southern part) were destroyed by the water, including aubergines, tomatoes, okra, etc. The plots high up on the dunes are not affected, but there are not many of them.

PhotolO: Market gardening areas located high up, north of Lake Wouye

Source: Cliché Kâ 2014

In 2018, only the higher plots to the north of the lake were not flooded. They are fewer in number than those to the south, and almost all of them are flooded. The majority of farmers (95%) have plots in the southern part of the lake. This represents a major loss for market gardening in the area.

- **Tanma Lake**

Around Lake Tanma, the potential agronomic fertility of the limestone and marl soils is conducive to market gardening. The main activity is market gardening, particularly in the southern, south-western and eastern parts of the lake, making Lake Tanma an important source of income. The lake's watershed was converted to market gardening 10 to 20 years ago. The main crops were potatoes, sweet potatoes and cabbage.

- **Situation in Taré**

At Taré, the fertility of the land made farming very pleasant. With easy access to water thanks to a shallow water table (around 2 to 3 m). The site is used for market gardening, as the land is ideal for market gardening. The southern, south-eastern and western parts of the site are used for market gardening. The majority of the village's inhabitants are farmers, who for years relied on market gardening for their livelihood. Onions dominate the crop, but other crops such as cabbage are also grown here.

- **Darou Mboumbaye**

Like Taré, market gardening plays an important role in the economic activities of the village of Darou Mboumbaye, with a wide range of crops including cabbage, onions and tomatoes. The opening of the breach in 2003 has had a negative impact on these crops, as salinisation continues to increase. The situation of the plots provides further clarification.

1.2 Location of plots in the study sites

- The case of Lake Wouye

In Dakar, most of the market garden plots on Lake Wouye were 1 ha in size, i.e. 10,000 m^2 (PLD, 2006). In 2007, with the problem of alkalinisation and salinisation, many plots had lost several square metres of their initial area. Some 51% of plots are between 1 ha and 3 ha in size. The 3 ha plots are numerous, belonging to a single family who farms them. There are also plots of between 2,040 and 5,000 m2 that have seen their cultivable area reduced by the advancing salt. But the plots most affected by this phenomenon are those between 100 and 1800 m2, which are the smallest, accounting for 29% and 10% of the total area.

Figure 16: Surface area of plots on Lake Wouye in 2006 and 2007

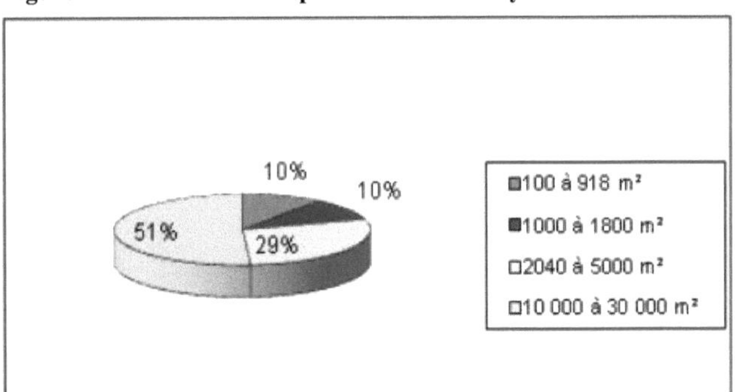

A lot of land has been lost due to alkalinisation of the soil and salinisation of the water, with repercussions on the income and living conditions of market gardeners. Surveys carried out in 2007 show that 50% of farmers have lost between 100 and 300 m2, and 35% between 300 and 400 m2. In 2010 and 2018, around 90% of market garden plots have been lost to the advancing waters of the lake.

Figure 17: Areas lost in 2007

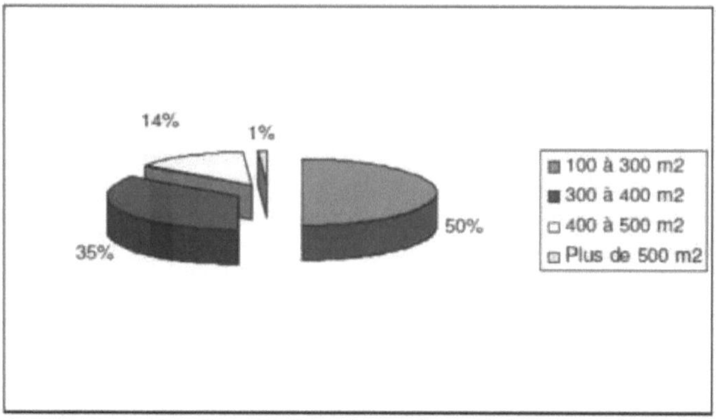

The consequences are reduced production, destruction of plantations, lower incomes and lower living standards. The most recent crops to be lost were carrots, cabbage, tomatoes, potatoes and aubergines. Around 50% of farmers grew cabbage and tomatoes before salinity and alkalinity. The farmers surveyed added potting soil (a mixture of soil and organic matter) to combat salinisation. Some added carbonate, mixed with manure. This explains the presence of $CO3^{2-}$ in some soil samples. This did not prevent the loss of 300 m^2 of arable land in some fields. This can be explained by the mineralisation of nitrogen in the presence of bicarbonate, which raises the soil temperature (it could heat up in the first few years) (Diop. 2007).

Lake Wouye lies close to the sea, separated by the Malika filao and coconut forest. This explains why the lake is naturally salty (EC of 10090 μS/cm in 2007), but its waters were drinkable in the 1960s (PLD, 2006). With evaporation, the salt is advancing towards the cultivable areas, causing farmers to retreat, and they are trying to combat salinisation by using chemical products (NPK fertilisers), but to no avail. After the lake was transformed into a retention basin in 2008, it would be difficult to give exact figures for the loss of arable land in the Lake Wouye area due to salinisation and alkalinisation. This represents hectares of land swallowed up by the waters of the lake, as a result of its transformation into a retention basin.

Figure 18: Situation of plots at Lake Wouye in 2018

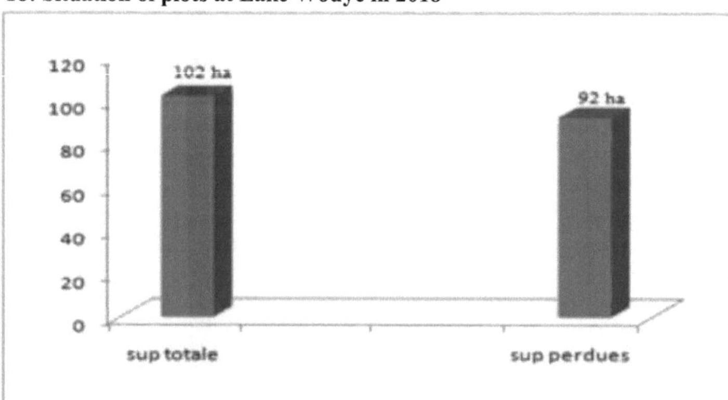

With a total surface area of 102 ha, the market garden areas of Lake Wouye have been divided into several plots of 1 ha each.

In 2018, a number of arable areas were flooded by the waters of the lake, resulting in the loss of around 92 hectares (90.2%). This is a disaster for market gardeners, for whom market gardening was their only source of income. The situation gets worse when the water level rises.

In 2007, the plots near the lake lost several areas to salinisation and alkalinisation.

In 2010 and 2018, around 90% of the plots around the lake were flooded. Analyses carried out on the lake show high water salinity (26,680 μS/cm in 2010 and 12,000 μS/cm in 2018), which will have a negative impact on the flooded plots. Ninety-six (96%) of the farmers surveyed in 2007 said that the soil was no longer producing as before, i.e. less than one tonne per ha.

In 2018, 93% of farmers who had lost their plot of land said that they were victims of injustice on the part of the State. Even though several areas were no longer suitable for farming, they had the option of selling the land for housing or poultry farming.

- **Tanma Lake**

Lac Tanma has 140 plots of 40 m$^{(2)\ each}$, covering an area of 5,600 m$^{(2).}$ Since 1988, market gardeners have found that every year they lose around 3 to 4 fields to salinisation and alkalinisation, making a total of more than 108 fields lost in 31 years (in 2019). The market gardeners who have lost their plots on Lac Tanma are cultivating the fields they have in the commune of Diender. The others, who have none, work as traders or apprentice drivers. Since 2010, salinisation and alkalinisation have been increasing. The solution recommended by farmers is to use organic manure (dried baobab leaves, cow dung, used by market gardeners who are close to the lake, because of the humidity). Poultry manure is used by those further away. Regular use of manure (every year) means that some land can be reclaimed (in the long term). In 2018, a plot of land taken near the lake in 2010, with cabbage plantations in poor condition, recovered almost half of its surface area, due to the discharge of wastewater from Sedima on part of the land. This is an experiment worth studying and analysing, as it could provide a solution to soil degradation.

Figure 19: Location of plots on Lac Tanma in 2018

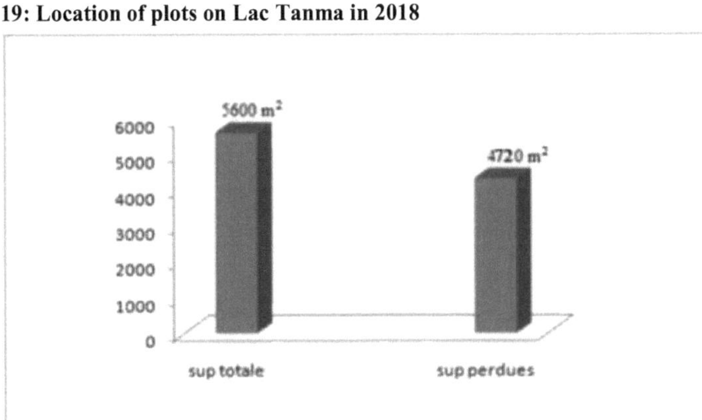

Covering a total area of 5,600 m2, with plots of 40 m2, Lac Tanma has lost a significant part of its cultivable area. Between 1988 and 2018, it lost around 4,720 m2. This is a huge loss for market gardening in the area, especially in economic terms. For 2019, the lake has lost 4,920 m2 of arable land (200 m2 more) than in 2018.

- **The Taré village website**

The village of Taré has plots measuring 1,000 m^2, with around 8 plots per concession (out of a total of 79). This gives an average of 8,000 m2 per family, with a total surface area of 63.2 ha (632,000 m2) in the village.

Figure 20: Situation of Taré plots in 2018

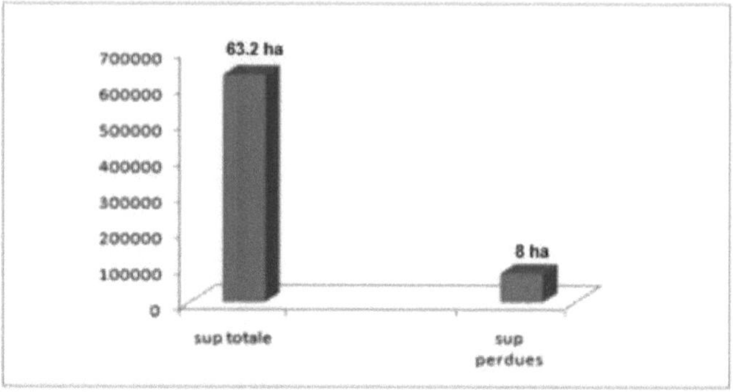

Of the 63.2 ha of arable land, around 8 ha is lost due to salinisation of the water and alkalinisation of the soil. No strategy has been put in place to combat this scourge. The only solution people have found is to farm new land in the sand dunes to the east of the village. However, as the salinisation of the water and alkalinisation of the soil increase each year, many plots become less profitable over the years.

The Darou Mboumbaye site

The Darou Mboumbaye site has plots of 1,000 m2, out of a total of 3,000 ha. Following the opening of the breach, some 2,250 ha were lost to salinisation of the water and alkalinisation of the soil. A huge loss for an area that relies heavily on market gardening.

Figure 21: Situation of Darou Mboumbaye plots in 2018

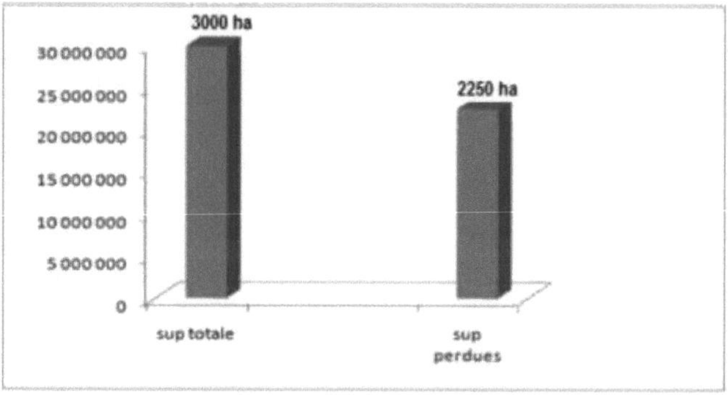

The market gardeners have formed an Economic Interest Group (EIG) and a Women's Promotion Group (GPF), with a total of 580 members in all the groups. Some of the group members have lost their plots of land. The only solution found is to rent land in the Potou sector and in Mboyenne, 7 km east of the village towards the Rao district. In 2018, the people who are no longer farming are fishing, working as farm labourers in Potou and Walo. Others are in commerce (opening shops), rearing goats, sheep, cattle or poultry. In 2019, the situation has not changed. It is important to know the salinisation situation in the different environments studied.

2. Salinisation in the study sites

The results reveal salinisation or alkalinisation of the soil and water in the study sites, confirming serious constraints. These chemical problems were noted in some of the wells in the market garden areas.

2.1. Results of soil and water analyses

Analysis of the samples taken has yielded great deal of interesting information.

2.1.1. Results of analyses of Lake Wouye 2010 and 2018

Electrical conductivity (EC) :

- Soils

In 2010, EC values ranged from 1 to 146.1 µs/cm, in 2007 they were 260 to 1,820 µs/cm, showing a decrease in 2010 and slightly saline to very saline soils in 2007. This situation is due to the transformation of the lake into a retention basin. Rising water levels have caused salinity to fall. The results of the analyses did not show the presence of salt in the soil in 2010, but rather alkalinity. This has had an impact on vegetable production. In 2010, we found aubergine plantations in poor condition, due to the salinity and the flooding of the market gardens.

Photo 11: Aubergine plantation found at Lake Wouye in 2010

Source: photo by M. Fall, September 2010

The condition of the plants shows traces of salinity on the leaves (dryness, dehydration, withering of the edges of the leaf blade), with necrosis on the leaves, which die and fall off. These are the first signs of plant degradation: disruption of the growth cycle. There is therefore a correlation between the plant's development and its salt content, which is higher in the leaves than in the stems. When it comes to growing aubergines, it should be noted that they are the least demanding of the Solanaceae plants in terms of soil type. Its powerful root system allows it to grow where other species have difficulty, but when it comes to planting, it can be more sensitive to soil type. Soils that are too wet are unsuitable. Similarly, soils with high salinity are unfavourable for this crop. In short, avoid poorly draining or asphyxiated soils. On the other hand, soils that warm well and are rich in organic matter and loam are favourable.

Figure 22: Soil EC of Lake Wouye in 2010

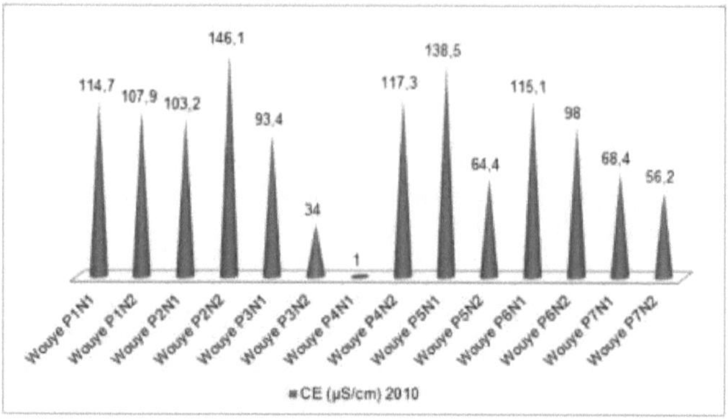

Photo 12: Plant dehydration and drought at Lake Wouye in 2010

Source: photos by M. Fall lac Wouye September 2010

Photos 1 and 2 show the dehydration and dryness of the plants around Lake Wouye in 2010. This was our first impression of soil degradation when we visited the site.

In 2018, electrical conductivity, shows slightly saline soils (Wouye P1N1, P1N2, P2N1, P3N1) to saline (Wouye P2N2, and P3N2), ranging from 290 to 670 µS/cm, with a higher EC at depth, showing an increase in soil salinity in 2018, compared to 2010 when there were no saline soils (1 to 146.1 µS/cm). This situation can be explained by the receding waters of the lake in 2018, which favours a rise in electrical conductivity.

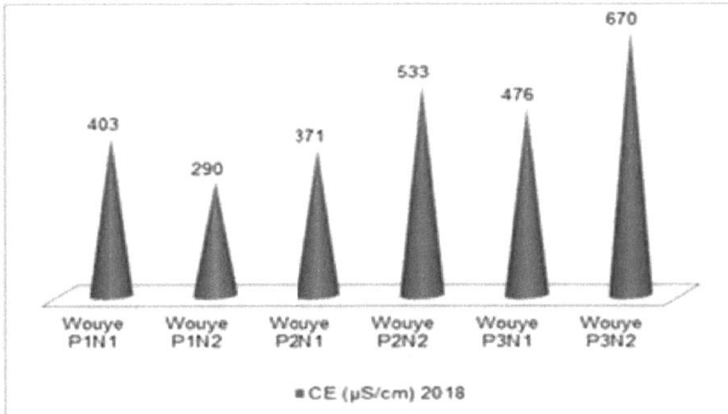

- Water

The results eutrophication of the lake's waters following its transformation into a retention basin wastewater from the flooded districts of Malika and Yeumbeul in 2008. The transformation of the lake into a retention basin led to flooding of the market gardens and the ceans in 2010, with saline water rising from the lake into the plantations and the ceans.

Photo 13: Flooding of market gardening areas on Lake Wouye in 2010

Source: Photo M. Fall September 2010

From left to right, water from the lake rushing into the market garden, with an unsuccessful attempt by the market gardeners to hold back the water with sandbags.

Please note: these results relate to samples taken after the lake was converted into a retention basin.

In 2008, a machine was installed to pump water from the lake to the sea to prevent flooding of the plots. The system does not seem to be working well in 2010, despite the presence of the machine, with almost all the plots under water. Analyses of the lake water showed high salinity, with an EC of 26,680 µS/cm, combined with eutrophication and a lack of oxygen (ammonium-rich water: $NH4^{+}$), making it difficult, if not impossible, for aquatic fauna to survive in the lake (disappearance of fish).

The lake water is not used directly to water the plants. In 2007, most of the plots had ceans, which have softer water and lower salinity, around 1696 µS/cm in 2007. The transformation of the lake into a retention basin in 2008 led to the intrusion of extremely salty water from the lake into the ceans, a risk to plant irrigation. The results of analyses on the lake showed an EC of 26,680 µS/cm in 2010, well normal (EC of 600 to 1,700 µS/cm), with a severe salinity level of more than 2,000 mg/l.

This excessive salinity is due to marine intrusions into the lake, because of its proximity to Atlantic Ocean. Irrigation water is pumped from canals and wells, causing a drop in water , leading to seawater intrusion into the shoreline. In 3 years (20072010), the salinity of the lake water has increased, with an EC of 10,090 µS/cm for the lake and 1,696 µS/cm for the ceanes, in 2007, compared with 26,680 µS/cm in 2010, for lake water. As the ceans were flooded, we were unable to take samples in 2010. Between 2007 and 2010, the lake's EC rose by 16,590 µS/cm (since it was converted into a wastewater retention basin). There is an increase of 5,530 µS/cm per year in the lake, making the water unsuitable for irrigation.

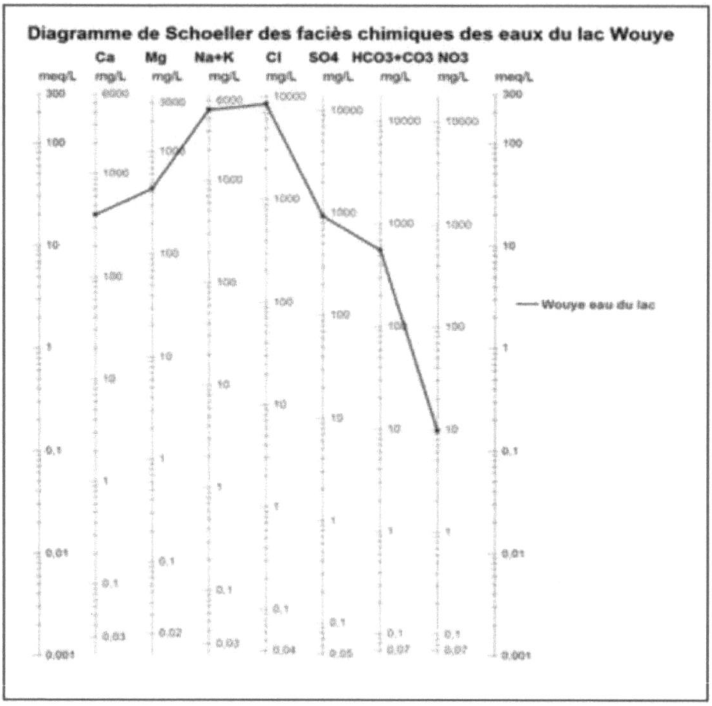

Figure 24 shows that the dominant chemical facies is sodium chloride, which indicates a state of salinity, with a dominance of chloride ions (8496.70 mg/l) and sodium (4715.69 mg/l), the main neutral salt dissolved seawater. The presence of the dominant seawater ions explains the maritime origin of the salts in Lake Wouye: marine intrusion. In 2010, the chemical facies showed a dominance of anions (9,971.16 mg/l), cations (5,697.98 mg/l) and nitrate ions (9.67 mg/l).

In 2018, the water in the ceans was very saline (1920 µS/cm) and the lake extremely saline (12,000 µS/cm). There was a drop in salinity compared with 2010 (26,680 µS/cm), with a difference of 14,680 µS/cm in 8 years, due to the influx water into the lake from the flooded areas of Malika and Yeumbeul.

It is important to note that in 2018, the lake has come back to life, with the return of fish and the resumption of fishing on the site due to the reduction in discharge of wastewater into the lake, unlike in 2010 when there was severe eutrophication of the waters and an almost total disappearance of fish. The waters of the lake, which flooded certain plots in 2010, have receded, freeing up arable land in the surrounding area. Market gardeners have tried to reclaim this land by adding organic matter, but the salinity level is too high and the harvests have not been good.

Figure 25: Salinity of Lake Wouye water in 2007, 2010 and 2018

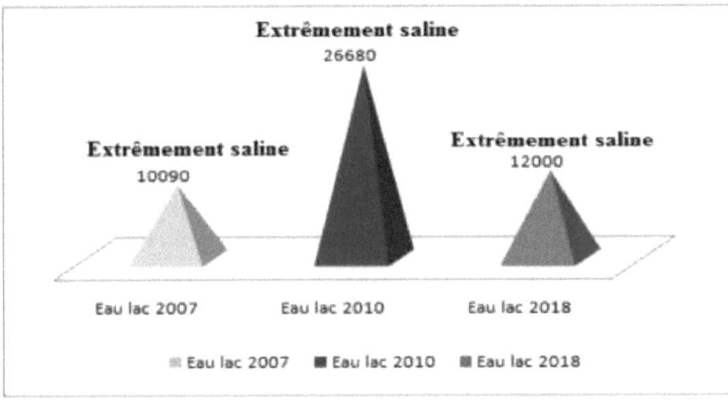

Figure 26 shows the chemical components of the water in Lake Wouye in 2018.

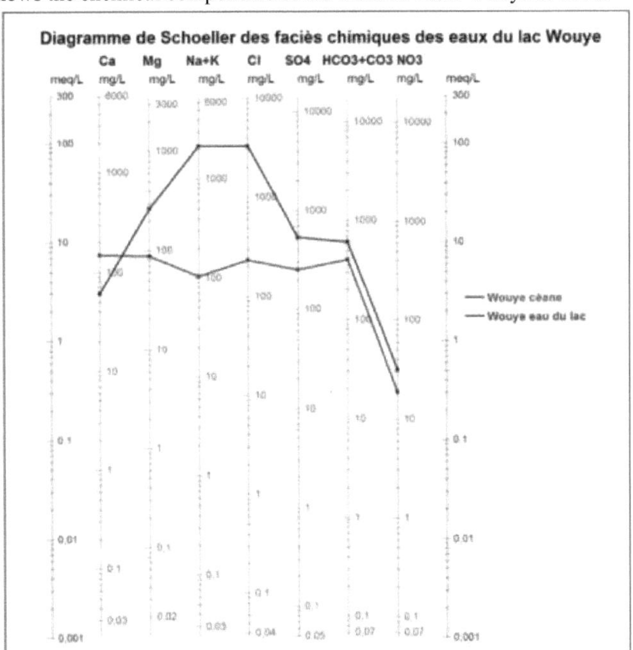

Sodium chloride (cl - 3308.64 mg/l and Na+ 2072.34 mg/l) is the main salt found in the waters of Lake Wouye, confirming the hypothesis of a marine intrusion. In the ceans, carbonate ions dominate (408.7 mg/l), followed by sodium (254.4 mg/l) and chloride (232.40 mg/l) ions. The chemical facies of the lake water show a predominance anions (5348.38 mg/l), compared with 2812.71 mg/l for cations. The presence nitrate ions (50.44 mg/l) indicates the use of chemicals in the environment.

- pH: soils

The pH values are more varied, probably due to the nature of the soils. They can range from neutral to acidic or alkaline, depending on the level of sampling. The nature of the soil can also vary over a short distance. The sampling points closest to the lake have neutral pH values varying between 6.6 and 7.2 (Wouye P1N1, P1N2, P2N1, P2N2 and P3N1). From P4N1 onwards, the pH changes, showing generally alkaline soils, ranging from 7.3 to 8.5.

This situation could be explained by the influx of water into the lake from flooded areas in the suburbs of Malika and Yeumbeul (pH of 6.90 in 2010), which causes a drop in the salinity level in this area. The further you move away from the lake, the lower the water table and the 74 marine intrusion, causing a rise in pH towards the surface (Wouye P4N1 and P7N1), with alkaline soils. In Wouye P5N2 and P6N2, the situation is reversed, with a higher pH level at depth than at the surface. Soils had a pH of between 5.18 and 8.30 in 2010, in 2007 it varied between 7.88 and 9.34, showing a drop in pH in 2010 due to the transformation of the lake into a retention basin.

The ideal soil is slightly acidic, with a pH of 6.5 (Wouye P3N1), which is suitable for most plantations. The variation in pH according to the points and levels of sampling is explained by the concentration of the soil solution. In soils of uneven moisture content, the concentration of the solution is inversely proportional to the moisture content. The transformation of the lake into a retention basin has created a high level of humidity in the nearest plots, hence drop in pH in these areas.

Figure 27: Soil pH in Lake Wouye in 2010

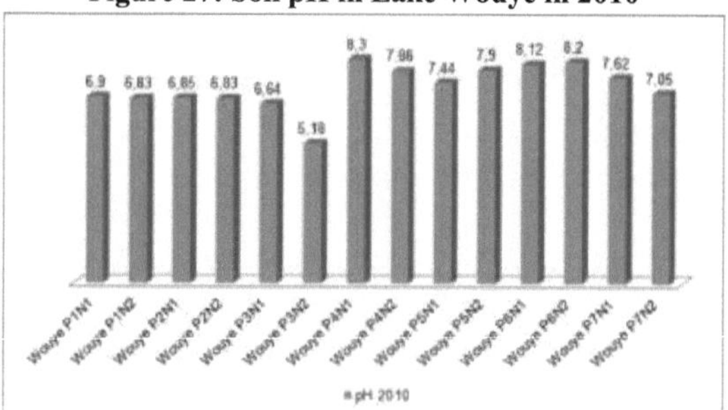

In 2018, the results show pH levels that are neutral at depth (Wouye P3N2), alkaline (Wouye P3N1) and very alkaline (Wouye P2N1 and P2N2), with higher alkalinity at the surface than at depth. The pH varied between 7.2 and 9.26 in 2018, much higher than in 2010 (5.18 and 8.30). This rise shows that the situation is worsening and approaching that of 2007, with a pH of 7.88 and 9.34. This confirms that it was the lake's waters that contributed to the drop in pH in 2010. In 2018, the waters had receded, certainly linked to the drop in rainfall (425 to 610 mm) in 2010, compared with (190 to 262 mm) in 2018 (ANACIM, 2018), which means that flooding has decreased, as has the quantity water discharged into the lake.

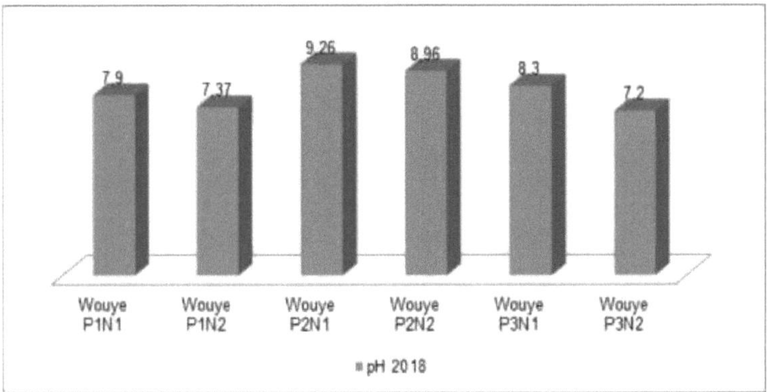

Analysis of the soil samples from Lake Wouye shows acidic and alkaline soils, not favourable to all crops. The cumulative results give an electrical conductivity of 13,240 µS/cm, indicating a highly saline area. So the real problem is how to develop market gardening in an alkaline and salty area.

- The pH of the water

The 2010 results show a neutral pH in the lake water (6.90), and slightly alkaline to alkaline in 2018, in the water of the ceanes (7.52) and the lake (8.28). There was an increase in pH of 1.38 between 2010 and 2018, which can be explained by the reduction in the amount of wastewater discharged into the lake in 2018 (less lake water flooding the market gardens).

Figure 29: pH of water in Lake Wouye in 2018

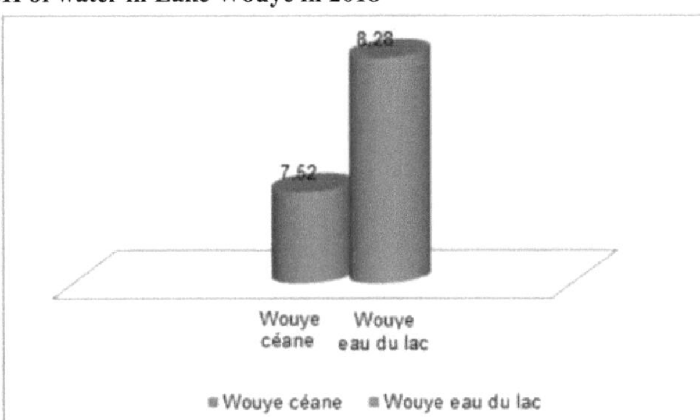

After Lake Wouye, the results of Lake Tanma in the Thiès region on the salinisation of the Niayes are presented.

2.1.2. Results of analyses of Lake Tanma 2010 and 2018

Electrical conductivity: soils. The EC values in 2010 remain almost the same on all the profiles, showing that at the spatial and point level (down to the 30 cm horizon), electrical conductivity is almost constant. This constancy is disrupted at Tanma P5N1 and N2, where there is an almost 4-fold increase between 0-15 cm and 15-30 cm depth. The 2010 results from Tanma lake show non-saline soils.

Figure 30: EC of Tanma Lake soils in 2010

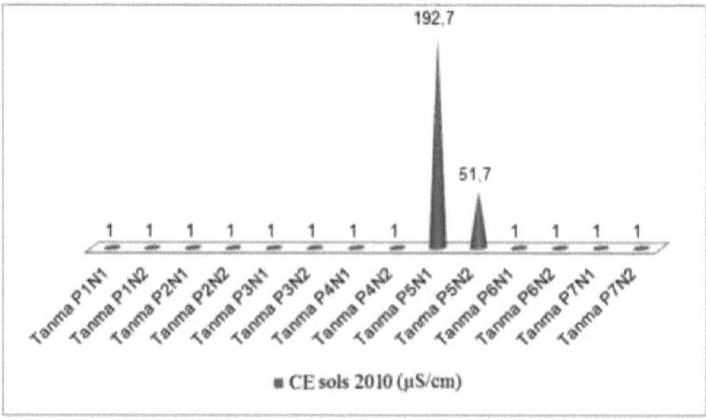

The electrical conductivity of Lake Tanma in 2018 shows a high salinity of the soils sampled on the surface of the lake (9630 µS/cm) due to marine intrusion (the northern part of the lake is close to the Atlantic Ocean). The results show non-saline soils in the vicinity of the lake (49 to 82 µS/cm), as for 2010 (1 to 192.7 µS/cm). Between 2010 and 2018, salinity increased. In 2010, almost all results were 1 µS/cm (Tanma P1N1, at P4N2 and P6N1 to P7N2) whereas in 2018, the data has evolved with a minimum of 49 µS/cm (Tanma P1N2). The 2018 results show a higher salinity at the surface (Tanma P1N1, P2N1 and P3N1) than at depth due to saline water irrigation coupled with high evaporative demand.

Figure 31: EC of Tanma lake soils in 2018

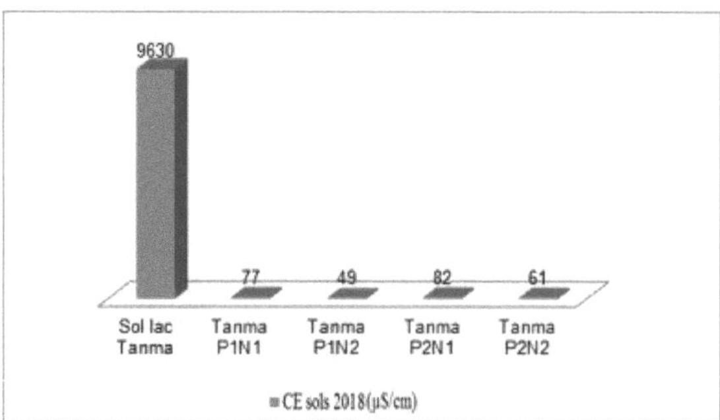

Water. irrigation water was highly saline in 2010, with an EC of 10,920 µS/cm. The high salinity of the irrigation water indicates marine intrusion into the environment and salinisation of the water table. Watering market gardening areas with such salty water can have a negative impact on yields.

Photo 14: Well at Lake Tanma in 2010

Source: photo by M. Fall October 2010

The images show irrigation wells photographed at Lake Tanma. Well 1 (image on the left) shows signs of eutrophication due to the use of chemicals. Well 2 (right) has traces of salinity in its surroundings, justifying the results of analyses showing a high level of salinity in the water (10,920 μS/cm).

Photo 15: Cabbage plantations at Tanma Lake in 2010

Source: photo by M. Fall October 2010

The two images show that plants have difficulty tolerating the presence of salts in water. Almost all the crops were parched, with wilting and dry leaves. Cabbage is one of the plants that tolerates salinity moderately well, so the condition of the plants shows that they have reached their tolerance threshold.

In 2018, two wells drilled in 2010 were abandoned. Samples were taken from the abandoned wells and from those still used for watering. The samples show a high presence of salts in the water, which is very saline (Tanma P1, 1500 μS/cm), to extremely saline (Tanma P2, 2200 μS/cm and P3, 7330 μS/cm). Tanma P2 is an abandoned well, while P3 is still in use, with a pump to facilitate watering. The higher EC level at Tanma P3 is due to water pumping, which can cause saline upwelling. The well taken in 2010 in the cultivated field (EC of 10920 μS/cm) has been abandoned, as has part of the plot that does not receive wastewater discharges from Sedima.

Figure 32: Chemical components of water taken from Lake Tanma in 2018

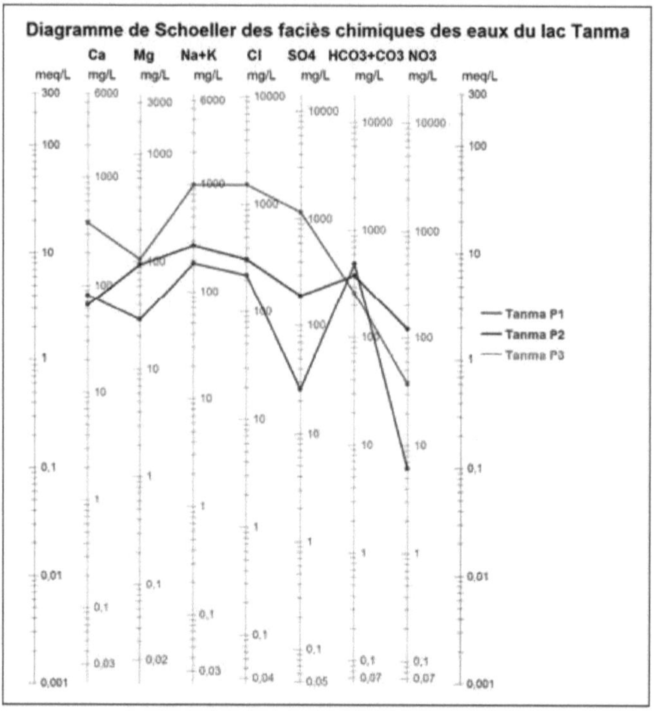

The diagram shows a strong presence of chloride and sodium ions (Tanma P3 and P2), with a dominance of bicarbonate ions in Tanma P1 (488 mg/l). Salts vary between 6.15 mg/l (Tanma P1) and 1547.73 mg/l (Tanma P3), with a strong presence of anions (Chloride, Sulphate, Hydrogen carbonate), 4588.82 mg/l in total and cations (Sodium, Potassium, Magnesium, Calcium), 2195 mg/l, present in seawater, which shows a marine intrusion in the environment. The results also show the presence of nitrate ions[6] (164.76 mg/l), revealing the use of chemical products in the environment.

Soil pH. The 2010 results show slightly alkaline to alkaline soils, ranging from 7.83 to 8.46. The pH values reveal a basic (alkaline) environment, with the presence of phosphate at Tanma P5N1 (192.5) and N2 (51.7). The presence of phosphate at Tanma P5 is due to the application acid fertilisers (mainly phosphate fertilisers) without tillage, leading to a concentration of acidity at the surface.

Excess phosphorus is generally not harmful to the crop. Phosphorus is one of the NPKs used extensively in the Niayes. Phosphoric acid (P) concentrates on the roots, ensuring their healthy development. It also promotes disease resistance and fertilisation (flowering, reproduction and fruiting). It is an important element for vegetables (tomatoes, aubergines, peppers, etc.), but to be effective, it needs to be incorporated into the soil.

---

[6] **Nitrate (NO_3^-)** A chemical fertiliser that is part of the nitrogen cycle.

Figure 33: Soil pH in Lake Tanma in 2010

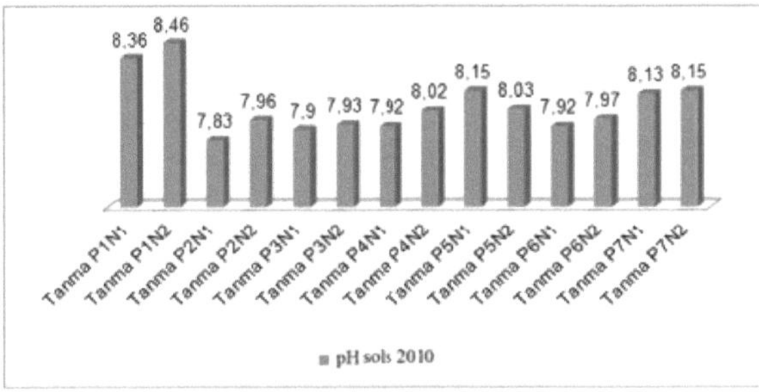

Photo 16: Soil conditions around Lake Tanma in 2010

Source: photo by M. Fall October 2010

In 2010, we solonetz (sodium-rich soil with an accumulation of clay in the **subsurface), which is more or less basic; plastic and viscous when wet (image 1, the one above), forming hard, compact clods when dry (images 2 and 3), it is asphyxiating, reducing and often sterile. We also noted the presence of limestone soil[7] all around the lake, in certain market gardening areas.**

Photo 17: Planting cabbages in limestone soil at Tanma Lake in 2010

Source: photo by M. Fall, October 2010

[7] **Limestone**: recognisable by the abundance of pebbles that constantly rise to the surface, whitish in colour and chalky in appearance.

Cabbage planted at Tanma lake in 2010, in chalky soil with pebbles (chalky) on the surface. Chalky soil has disadvantages because it retains fertilising elements, causing chlorosis[8] of the leaves (the active chalk hinders the plants' uptake of iron).

Photo 18: Sweet potato plantation with leaf chlorosis at Lake Tanma in 2010

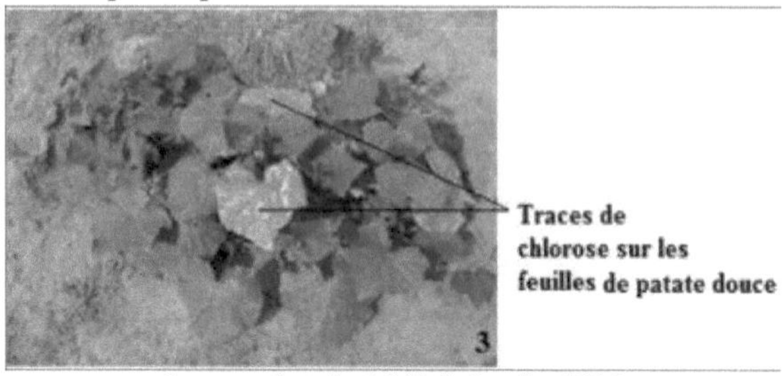

Source: photo by M. Fall, October 2010

Chlorosis of the leaves appearing on the sweet potato plantation, showing a retention of

fertilising elements in the soil. As sweet potato plantations are not grown in calcareous soil, we can deduce that watering with water containing an inadequate level of active limestone is necessary.

Several vegetables are adapted to limestone soil: lettuce, celery, potato, cabbage, radish, aubergine, shallot, tomato, carrot, pepper, onion, pea (jardiland 2014, X. Gerbeaud 2016).

In 2018, the pH varies between 8.23 and 9.32, with alkaline to very alkaline soils, higher than in 2010 (7.83 to 8.46). Alkalinity is higher at depth (Tanma P1N2, P2N2) than at the surface (Tanma P1N1, P2N1). In contrast to 2010, we took soil samples from the surface of the dried lake, showing soil alkalinity (8.23).

Figure 34: Soil pH in Lake Tanma in 2018

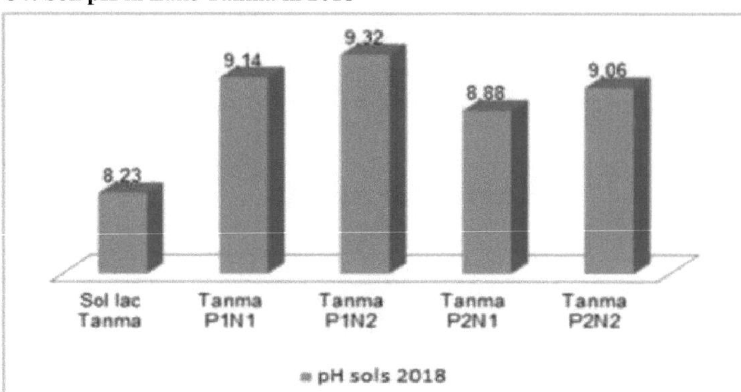

- Water

The 2010 results show a neutral pH, ranging from 6.7 to 7.2. In 2018, the results show alkaline water, ranging from 7.63 to 8.21, with higher alkalinity in Tanma P1, which represents Sedima's wastewater (organic matter, made up of a mixture of poultry waste

---

7**Chlorosis:** a physiological plant disease in which the leaves become discoloured due to problems with chlorophyll production, linked to a lack of nutrients.

"excrement, blood", with water). This is the water recovered after cleaning the hen houses. Sedima, in collaboration with certain market gardeners, discharges the waste water into the fields for fertilisation. This initiative is producing satisfactory results in terms of recovering degraded land. One market gardener claims to have seen an increase in production since the water was poured onto his plantations. The cultivated plot tested in 2010 confirms the effectiveness of discharging Sedima's wastewater onto the fields. The part of its surface area that receives the water is more fertile (presence of plantations) than the other part, which was abandoned (uncultivable). Compared with 2010, the pH level increased in 2018 from 0.93 to 1.03, showing a change in the degradation of the irrigation water over the years. Our interviews confirm this state of affairs, with market gardeners saying that the water is no longer of good quality for watering.

Photo 19: Photographs of Lake Tanma in 2018

Source: photo by M. Fall lac Tanma July 2018

Photo 1 shows the Sedima truck discharging wastewater (Tanma P1) onto the plot (photo 2), with tomato plantations. Photos 3 and 4 show the well being pumped to water the cultivated field (Tanma P3). Photo 5 shows the well abandoned in the cultivated field (Tanma P2) because of poor water quality. It should be noted that the well still in use (Tanma P3) is not far from the abandoned well (Tanma P2). This could explain why Tanma P3 has a higher salinity level, due to the pumping of water, which causes salinity to rise. Photos 6, 7 and 8 show the field cultivated in 2010, where soil and water samples were taken. In 2018, part of the field has been abandoned (photo 8), but another part is still being farmed (photos 6 and 7). The exploited part receives wastewater discharges from Sedima, which has made it possible to reclaim land.

The results from the Louga region will provide more information on salinisation in the Niayes area, especially in the northern part.

2.1.3. Results of Taré analyses 2010 and 2018

Electrical conductivity

- **Soils**

The results of the 2010 samples show non-saline soils, with electrical conductivity ranging from 27.7 µS/cm (Taré P1N1) to 52.5 µS/cm (Taré P7N1). They show that electrical conductivity is higher at the surface than at depth (Taré P2N1, P3N1, P6N1 and P7N1), due to irrigation with saline water and evaporation phenomena that favour the accumulation of salts at the surface. Compared with Dakar, EC levels are lower.

Figure35: EC of Taré soils in 2010

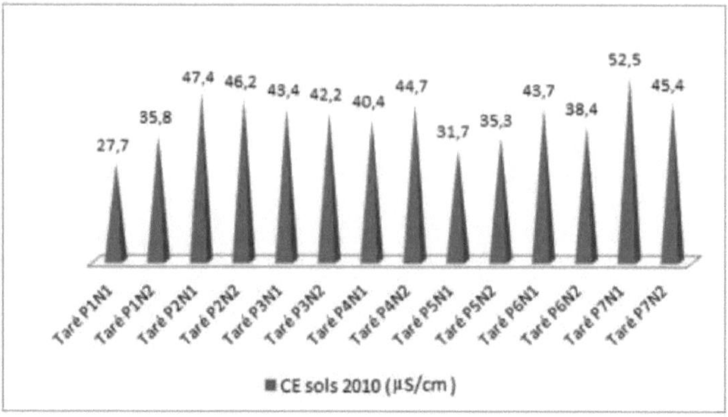

In 2018, the electrical conductivity shows non-saline soils, with an EC of 32 µS/cm to 86 µS/cm. Compared to 2010 (27.7 µS/cm to 52.5 µS/cm), with an increase of 4.3 µS/cm to 33.5 µS/cm in 8 years. This shows an increase in salinity in the environment, with a higher salt level at the surface than at depth (Taré P1N1, P3N1).

Figure 36: EC of Taré soils in 2018

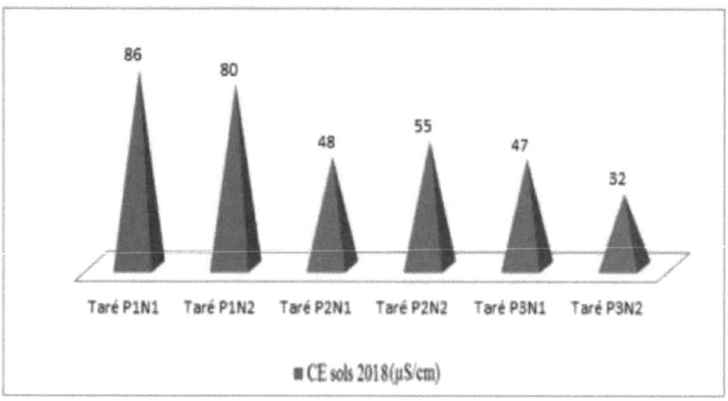

- **Water**

The 2010 results show non-saline water, with an EC of 1 µS/cm. This means that in 2010, alkalinity was the problem in the market gardening areas of the village of Taré, with crops not adapted to alkaline soils and alkaline water. In 2018 (2,120 to 45,000 µS/cm), the results show extremely saline water, much higher than in 2010 (1 µS/cm); in 8 years, the salinity of the well water and river water in Taré has largely increased due to marine intrusion.

This situation can be explained by the opening of the breach on the Langue de Barbarie in

2003, which has been growing ever since. From 2010 to 2018, we have an increase in EC that varies from 2,119 to 44,999 µS/cm (Taré P1, 3,000 µS/cm, Taré P2, 2,120 µS/cm, Taré P3, 19,000 µS/cm), with a higher level in the Senegal River, which varies from 1 µS/cm in 2,010 to 45,000 µS/cm in 2018, which is really high, going from non-saline to extremely saline in 8 years. Taré P3 (19,000 gSZ cm) is closer to the river (around 100 m), showing marine intrusion from the river, with the breccia opening in 2003, progressing from 750 m in 2004, to around 7 km in 2018.

Figure 37: Chemical components of well and river water collected at Taré in 2018

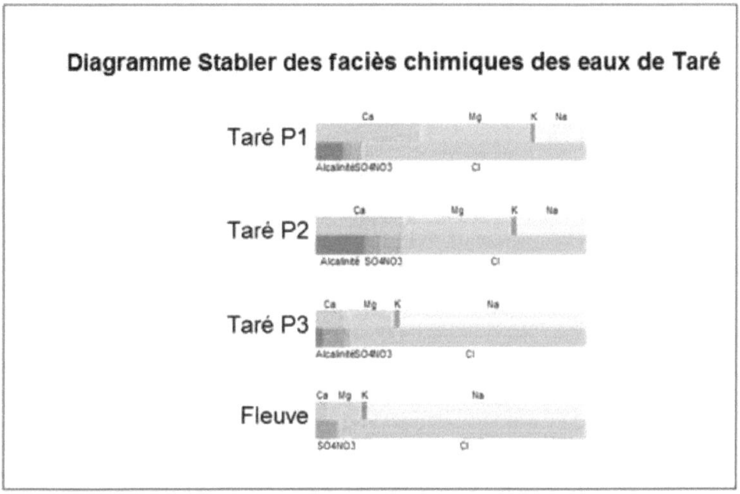

The chemical facies of the water show a dominance of chloride and sodium ions in the wells and in the river. There is a higher salinity in the river, with the presence chloride (13,595.09 mg/l) and sodium (8,034.18 mg/l) ions, showing marine intrusion into the river, reinforced by the opening of the breach in 2003. The high presence chloride (5,787.93 mg/l) and sodium (3,083.75 mg/l) in Taré P3 indicates marine intrusion from the river. The ions present in Taré water samples show the presence anions (chloride, sulphate, carbonate) and cations (sodium, potassium, calcium), which are found in seawater, confirming marine intrusion into the market garden areas.

Soil pH

2010 results show alkaline pH values in all soil samples, ranging from 7.83 to 8.62. They are higher at depth than at the surface (Tare P2N2, P3N2, P6N2 and P7N2), which can be explained by the marine intrusion into the environment.

Figure 38: pH of Taré soils in 2010

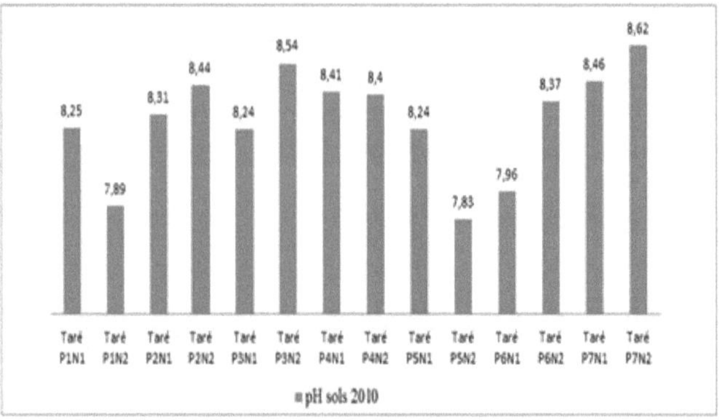

Unlike 2010, with pH values ranging from 7.83 to 8.62, the 2018 results are higher, ranging from 8.62 to 8.96. They show very alkaline soils in the village of Taré, with higher alkalinity at depth than at the surface (marine intrusion), which changes over the years. This change has an impact on market gardening (not all crops adapt to alkaline soils). In 8 years, there has been an increase from 0.34 to 0.79. A slow increase, but one that could significantly affect yields. From 2010 to 2018, the pH has become very alkaline in the village of Taré.

Figure 39: pH of Taré soils in 2018

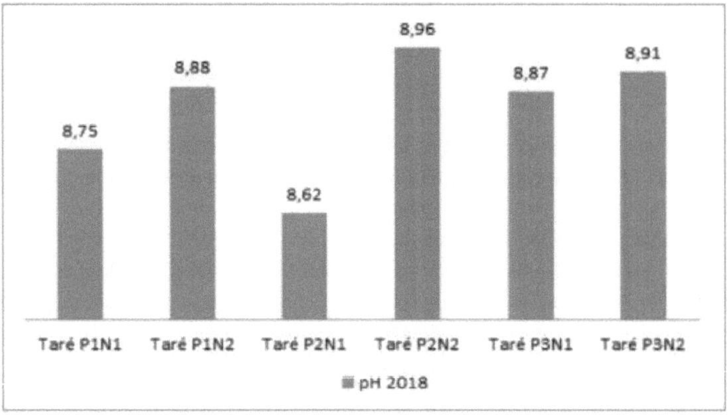

- Water

The 2010 results show alkaline water, ranging from 7.96 to 8.32. 2018 results show slightly alkaline water (7.25 to 7.78), with a drop compared with 2010 (7.96 to 8.32), due to the fact that the market gardeners have abandoned the plots where samples were taken in 2010. One well sampled in 2010 in a cultivated field was sampled in 2018. It is the abandoned well closest to the market gardening areas still under cultivation in 2018 (Taré P3). The well has a higher pH than the others (7.78), justifying the retreat of market gardening inland and the drop in pH. Of the plots sampled in 2010 and abandoned in 2018, we chose the plot closest to the market garden areas that are still suitable for cultivation. The higher pH level in the Taré P3 well shows a shift in alkalinity inland (Taré P2, 7.32, Taré P1, 7.25). The results from Saint-Louis show a trend towards salinisation in the *Niayes*, particularly to the north.

2.1.4. Darou Mboumbaye analysis results 2010 and 2018

The electrical conductivity of soils

EC results in 2010 ranged from 1 to 173.1 µS/cm, revealing non-saline soils, with an accumulation of salts at the surface (Darou Mb. P4N1, Darou Mb. P5N1, Darou Mb. P6N1 and Darou Mb. P7N1), due to marine intrusion and saline upwelling towards the surface. At Darou Mb. P3N1, there is a large difference between 0 - 15 cm and 15 - 30 cm, varying from 1 µS/cm (0 - 15 cm) to 102 µS/cm at depth (15 - 30 cm), due to saline upwelling from the water table. The results show that the first samples taken (Darou Mb. P1N1, Darou Mb. P2N1 and Darou Mb. P3N1) have a higher EC at depth than at the surface, which explains an inland trend in salinity, starting from the village of Darou Mboumbaye.

Figure 40: Soil EC in Darou Mboumbaye in 2010

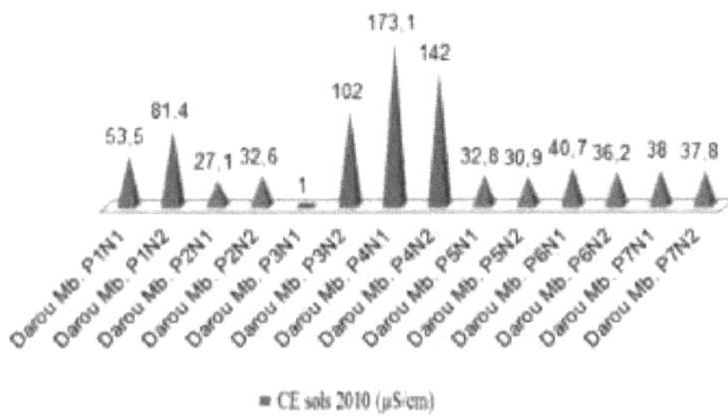

The 2018 results show non-saline soils as for 2010, with an electrical conductivity varying between 48 µS/cm and 124 µS/cm. Compared with 2010 (1 to 173.1 µS/cm), we have higher EC at the surface than at depth (Darou Mb. P2N1, Darou Mb. P3N1). We noted a higher electrical conductivity in the field still under cultivation than in those abandoned, with a higher level at the surface (Darou Mb. P2N1, 124 µS/cm). This can be explained by watering with high salinity water. The values for Darou Mb. P3N1 (55 µS/cm), an abandoned field (which is no longer watered) confirm this idea, with a lower EC in the field still under cultivation. In the abandoned plot just behind the village (Darou Mb. P1N1 and N2), the EC is higher at depth (115 µS/cm) than at the surface (88 gSZ cm), showing saline upwelling towards the surface.

Figure 41: Soil EC in Darou Mboumbaye in 2018

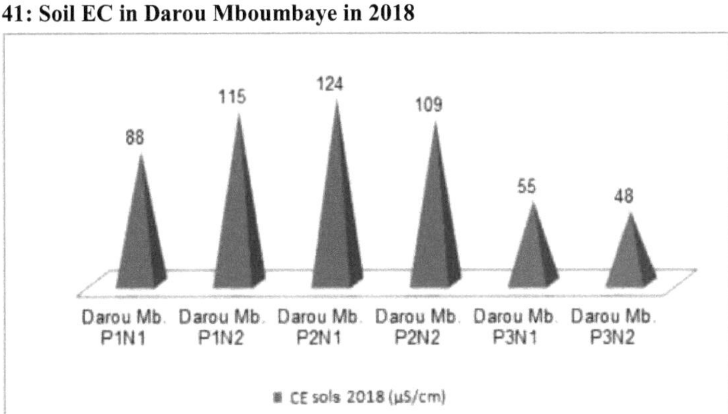

Water

The 2010 results show non-saline water ranging from 30.4 to 30.5 µSZcm, with a higher level in the well of the field still under cultivation (Darou Mboum P1). This shows salt rising to the surface due to watering. The results for 2018 are much higher than those for 2010. The EC shows extremely saline water, ranging from 9,800 µSZcm (Darou Mboum P3) to 18,200 µSZcm (Darou Mboum P1), with a high presence of chlorine in the water, showing marine intrusion into the water table. In 8 years, the situation has evolved in a critical manner, with a sharp increase in the salinity of the water.

Figure 42: Chemical components of well water in Darou Mboumbaye in 2018

Stabler diagram of the chemical facies of Darou Mboumbaye waters

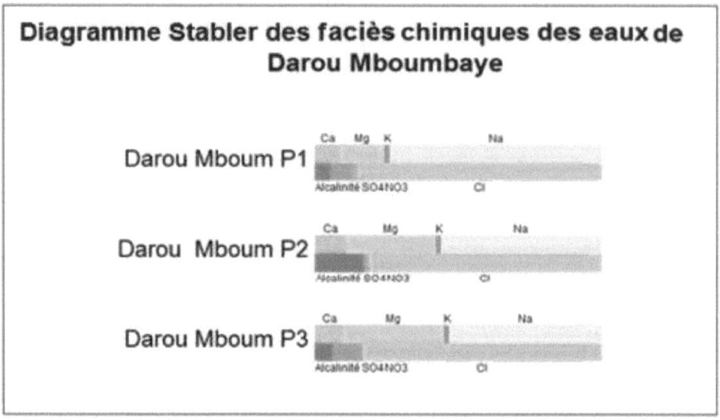

The chemical facies of the Darou Mboumbaye water show a strong presence of chloride ions (2856.29 to 5393.40) and sodium ions (1162.48 to 3028.02) in the wells, with a greater quantity of chloride ions, due to marine intrusion, aggravated by the opening of the breach in 2003. The quantity of ions is higher in the first wells (Darou Mboum. P1 and P2), with respectively 5,393.40 and 4,132.17 for chloride and 3,028.02 and 1,851.4 for sodium, due to a shift in salinity from the village inland (Darou Mboum. P1 is about 15 m from the village, in an abandoned field).

The pH of Darou Mboumbaye soil

The pH in 2010 was almost the same as at Taré (alkaline soils), except that here all the values were above 8 (8.05 to 8.95). Soil acidity varies according to the sampling points and depths. In Darou Mb. P1N2 (8.95), P2N2 (8.39) and P5N2 (8.26), alkalinity is higher at depth than at the surface. At Darou Mb. P3N1 (8.31), P4N1 (8.64), P6N1 (8.38) and P7N1 (8.65), the situation is reversed, with a higher alkalinity at the surface than at depth, showing an evolution in alkalinisation.

Figure 43: Soil pH in Darou Mboumbaye in 2010

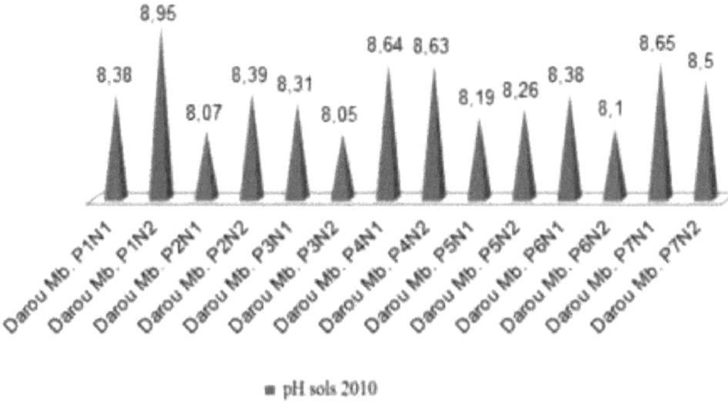

The 2018 results show basic pH levels (8.87 to 9.34), with alkaline soils that are not suitable for many crops. Like Taré, all the soil samples have high pH levels, but those from Darou Mboumbaye are higher. Compared with 2010, the pH level has increased, showing a change in soil acidity over the years. The increase in pH can explained by the salinity of the irrigation water and climate change, with irregular rainfall. Humidity lowers the pH level by making it acidic or neutral. If there is less water in the soil, the pH is likely to rise and become basic. Chemical fertilisers increase acidity. Their nitrification or absorption leads to the production of protons .[9]

Figure 44: Soil pH in Darou Mboumbaye in 2018

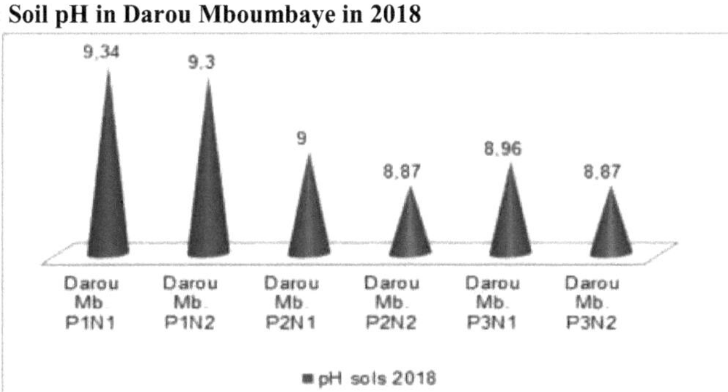

- The waters of Darou Mboumbaye

The wells used in Darou Mboumbaye in 2010 showed alkaline water (> 7), ranging from 7.26 to 7.97 in the cultivated and uncultivated fields, with a higher pH in the well in the abandoned field (Darou Mboum P2), due to irregular rainfall and a drop in water supply. The 2018 results are better, with pH values ranging from 8.05 to 8.19, showing a change in alkalinity over the years, from 0.22 to 0.76. The alkalinity of the water worsens over the years. A summary of the results will shed more light on changes in pH and salinisation in the *Niayes*.

3. Summary of results

3.1. Changes in pH and EC in the *Niayes* in 2010 and 2018

In 2010, the pH results showed alkaline soils with variations in levels depending on the sampling points. Soil alkalinity was higher at Darou Mboumbaye (8.05 to 8.95) and lower at

8The **proton**: is a subatomic particle "which refers to particles or distances smaller than those of atoms" carrying a positive elementary charge.

Lake Wouye (5.18 to 8.30). In 2018, there was an increase in pH in almost all the environments studied. From 5.18 to 8.95 in 2010, the pH of the soils at the four sites rose from 7.2 to 9.34. It should be noted that in 2010, the pH showed acidic, slightly alkaline, alkaline and very alkaline soils. The 2018 results show a change in the situation, with a pH of neutral (Wouye P3N2), alkaline and very alkaline soils.
The water was alkaline overall in 2010, with high alkalinity in the wells in the village of Taré (8.23 to 8.32). The pH increased from 6.90 to 8.32 in 2010 and from 7.25 to 8.30 in 2018. The pH in 2010 showed water that was neutral (Tanma P1), slightly alkaline and alkaline. The 2018 results show a change in the situation, with the pH of the water alkaline.

Figure 45: Changes in the pH of water at different sites in 2010 and 2018

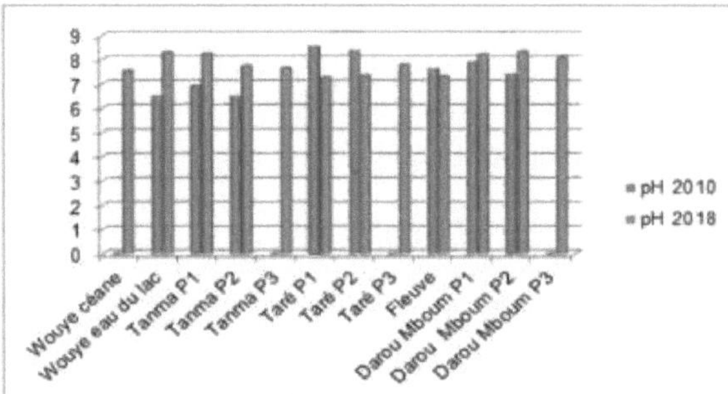

The *Niayes* EC results showed non-saline soils in 2010 ranging from 1 to 192.7 gSZ cm. In 2018, there was an increase in EC at almost all the sites studied, ranging from 32 to 9,630 µSZcm, with soils that were non-saline (Lake Tanma, Taré and Darou Mboumbaye fields), slightly saline and saline (at Lake Wouye) to extremely saline (dried-up Lake Tanma soil).

Figure 46: Changes in soil EC at different sites in 2010 and 2018

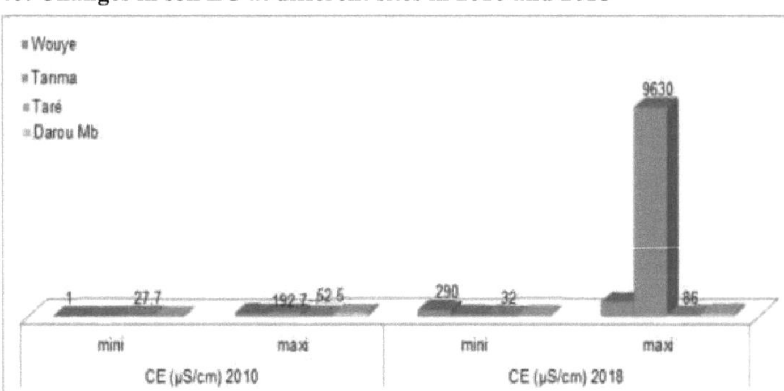

Overall, the water is very saline in the water in Lake Wouye (26,680 µSZcm) and in a well in Lake Tanma (10,920 µSZcm) in 2010. We observe an EC that increases from 1 to 26,680 µSZcm in 2010, and from 1,500 to 45,000 µSZcm in 2018. The EC shows non-saline water at Taré and Darou Mboumbaye, and extremely saline water at lakes Wouye and Tanma in 2010 and very saline water (Lake Wouye, Tanma P1), extremely saline water (Lake Wouye, Tanma P3, Taré P2 and P3, Senegal River, Darou Mboum P1 and P3) in 2018.

Figure 47: Variation in EC of water at different sites in 2010 and 2018

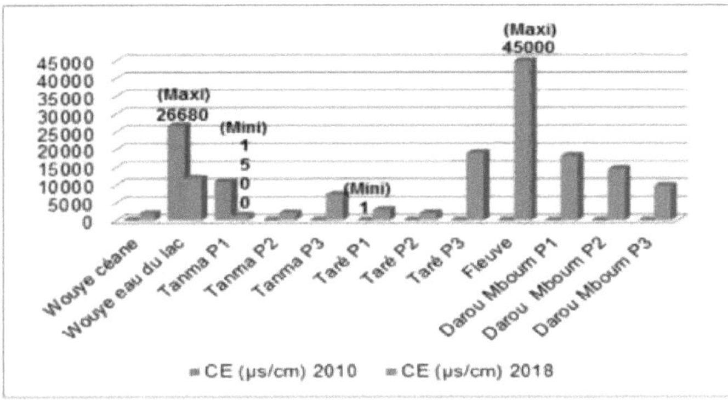

3.2. Changes in salinity in the *Niayes* in 2010 and 2018

The results obtained in 2018 show changes in salinisation in the *Niayes* between 2010 and 2018 (8 years). Most of the values are higher than in 2010 at all the sites studied. To get an overview, we calculated the average EC for each site so that we could make a comparison between them and get a clearer idea of how salinisation is changing in the *Niayes*. The average results showed an increase in salinity in the *Niayes* from 2010 to 2018.

In 2010, the average results showed greater salinity around Lake Wouye, with an average of 85.71, and a total EC of 1,200 µS/cm, followed by Darou Mboumbaye 59.22 (Total EC, 829.1 gS/cm), Taré 41.05 (Total EC 574.8 µS/cm) and Lake Tanma 18.31 (Total EC 256.4 µS/cm). Cumulative ECs show slightly saline, saline and very saline soils in 2010.

In 2018, the average results show a higher salinity at Lake Tanma, due to the high salinity of the soils in the dried lake (9630 µS/cm), with an average of 1,649.83 and a total EC of 9899 µS/cm. Lake Tanma is followed by Lake Wouye, with an average of 457.16 (total EC 2743 µS/cm), Darou Mboumbaye 89.83 (total EC 539 µS/cm) and Taré 58 (total EC 348 µS/cm). Cumulative ECs for 2018 show slightly saline, saline and extremely saline soils.

Figure 48: Average soil EC values for the various sites in 2010 and 2018

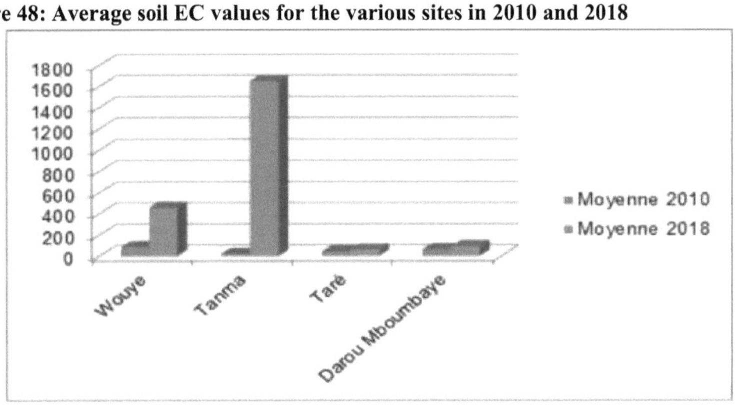

In 2010, the water at Taré and Darou Mboumbaye was not saline (between 1 and 30.5 µS/cm), but extremely saline at Lake Wouye and Lake Tanma (from 10,920 to 2,680 µS/cm). In 2018, the waters are very saline in Lake Wouye (lake ceane, 1920 µS/cm) and Lake Tanma (Tanma P1, 1500 µS/cm) and extremely saline in the waters of Lake Wouye (12000 µS/cm),

Lake Tanma (Tanma P2, P3), Taré (Taré P1, P2 and P3), Darou Mboumbaye (Darou Mboum 1, 2 and 3) and the Senegal River (45000 µS/cm).
The average EC in 2010, shows a higher salinity of water in Lake Wouye, 26680, followed by Lake Tanma (10,920) and Darou Mboumbaye (30.45). In 2018, water salinity was highest in Darou Mboumbaye, with an average of 14,166.66, and a total EC of 42,500 µS/cm, followed by Taré (8,040), Lake Wouye (6,920) and Lake Tanma (3,676.66). This shows that salinisation is a recurring problem in the Niayes, but especially salinisation of the water. This situation is indicative of strong marine intrusion in the area, encouraged by climate change (global warming, coastal erosion, reduced rainfall, drought, salinisation, etc.). If we take into account the Senegal River in the northern *Niayes* area, water salinity is highest in the river (45,000 µS/cm) in 2018, followed by Darou Mboumbaye with a total EC of 42,500 µS/cm, Taré (24,120 µS/cm), Lake Wouye (13,920 gSZ cm) and Tanma (11,030 µSZcm). The presence of the breccia and irregular rainfall in the northern zone explain this situation.

Figure 49: Average EC values for water at different sites in 2010 and 2018

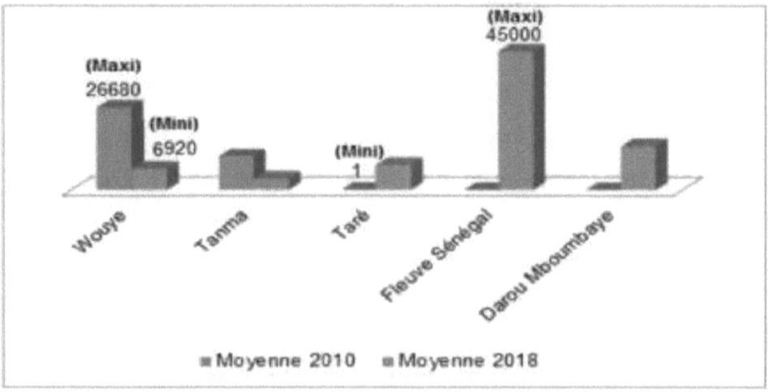

Between 2010 and 2018, the ECs in the study environments have undergone varying changes. At soil level, Lake Tanma has a change varying between 48 and 9,437.3 µSZcm, followed by Lake Wouye with a change from 289 to 523.9 µSZcm, the village of Taré 4.3 to 33.5 µSZcm. Darou Mboumbaye saw an increase in the minimum value of 47 µSZcm and a decrease in the maximum value of 49.1 µSZcm, due to the use a less salty well in 2018, the 2010 one is abandoned.
The results of the wells sampled on Darou Mboumbaye P4N1 in 2010 and P2N1 in 2018), show a higher salinity of the abandoned well (14,500 µS/cm), used in 2010, while the well used in 2018 is 9,800 µS/cm.

Figure 50: Changes in soil EC at the study sites in 2010 and 2018

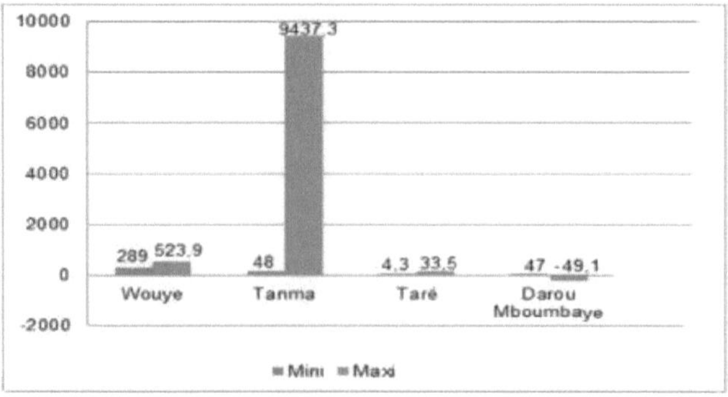

In 8 years (2010-2018) the village of Darou Mboumbaye has seen a sharp change in water EC, with an increase varying between 9,769 and 18,169.5 µS/cm, followed by Taré between 2,120 and 18,999 µS/cm, Lake Wouye 1,920 to 14,680 µS/cm and Lake Tanma 1,500 to 3,590 µS/cm. The high salinity of the Darou Mboumbaye and Taré wells shows the extent of marine intrusion into *the Niayes*, encouraged by the breccia.

Figure 51: Changes in EC water levels at the study sites in 2010 and 2018

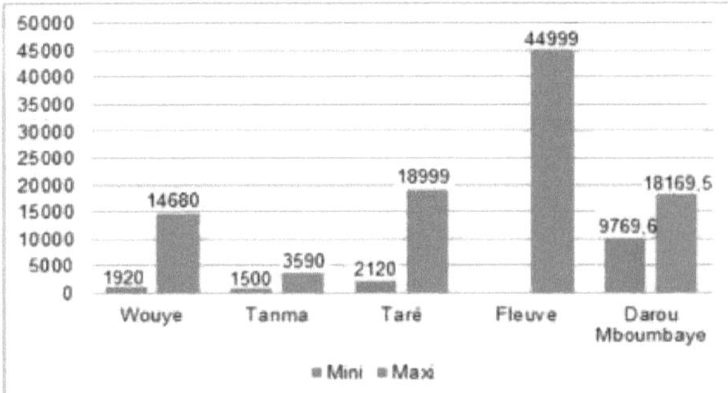

Conclusion

Chapter 3 presents the results obtained. The market gardening activities of the study sites are faced with several difficulties related to salinisation. The results of *in situ* surveys provided information on the various crops, the situation of the plots and yields. The results of the analyses showed the salinisation situation in the *Niayes*, which has consequences for market gardening. Salinisation has evolved over the years, with a higher level in the northern part of the *Niayes*, due to the breach, which increases marine intrusion. In Lake Wouye, there is an accumulation of salts on the surface and at depth, with high water salinity. pH levels are lower at Lake Wouye than at the other sites.

At Lake Tanma, the salinity of the soil is almost constant and the water is not suitable for irrigation because the salinity is too high.

Taré has a higher soil EC at the surface than at depth, and its water was alkaline in 2010 and extremely saline in 2018. The Senegal River goes from alkaline water to extremely saline water in 8 years (2010-2018).

Darou Mboumbaye has alkaline soil, with the highest pH levels of the four regions, alkaline water that was not favourable to all crops in 2010 and extremely saline in 2018. On the whole, we have alkaline soils in 2010 and in 2018, except in the case of the soil at Lake Tanma, which has dried up in the Thiès region, with saline soils in 2018.

The results showed saline water throughout the *Niayes*, with higher salinity in the north, in the Saint-Louis and Louga regions.

In Dakar, the transformation of Lake Wouye into a retention basin lowered the pH level at certain points in 2010, making the soil neutral or acidic. In 2018, as the water recedes, the pH level rises, making the soil alkaline. Flooding of the plots makes them unfit for cultivation, as the lake water is highly saline (due to marine intrusion). The analysis of the results and their interpretation in chapter 4 will enable the results set out in chapter 5 to be discussed.

Chapter 5

Analysis, interpretation and discussion of results

The *Niayes* area is faced with demographic problems and heavy urbanisation. Urban and peri-urban agriculture is practised in the area thanks to favourable soil, hydrological and climatic conditions. In this area, the water table is very shallow, between 1 and 3 metres depending on the location and the season. The use of water from the water table or *Céanes*[10] to irrigate fields has led to the intrusion of saline water, which has reduced the quality of the water. Years of drought and shifting isohyets in recent years have led to a reduction in the rainfall on which agriculture, the main source of the national economy, depends.

This situation has repercussions on the lowering of the water table, leading to the intrusion of salt water. It is important to note that, apart from natural causes such as drought and coastal erosion as the sea advances, man plays a major role in the salinisation process *in the Niayes* area. Excessive exploitation of the water table (pumping of water), excessive use of chemical products (pesticides) and the opening of the breach south of Saint-Louis have resulted in an advance of the salt wedge towards the market garden areas, causing salinisation (increase in chlorides and sulphates) and alkalinisation (increase in bicarbonates), which have a number of impacts on market garden activities. This situation of salinisation and alkalinisation of the soil and water is not without consequences for the plantations.

Salt disrupts plant development by limiting the uptake of nutrients and reducing the quality of water available to plants. High soil salinity levels cause plants to wither as a result of increased osmotic pressure and the toxic effects of salts. Excess sodium leads to the destruction of the soil structure, which is unable to support plant growth due to a lack of oxygen. Salinisation increases the impermeability of the deep layers of the soil, and the soil can no longer be used for all crops. Interpretation and discussion of the results will shed more light on the salinisation of our study sites.

1. Interpretation and discussion of analysis results

The results of the sample analysis show that the soil is alkaline, which has a negative impact on market gardening.

1.1. Location of study sites

All four sites have non-saline soils, except for the soils sampled at Lake Wouye in 2018 (Wouye P2N2 and P3N2), with saline soils (533 μS/cm and 670 μS/cm) and on the surface of dried Lake Tanma, which are extremely saline (9,630 μS/cm).

For Durand J. H. (1983), from 0 to 500 μS/cm we have non-salty soils that have a negligible effect on yield, whereas from 500 to 1000 μS/cm, we have slightly salty soils that reduce yields of very salt-sensitive crops, such as Wouye P2N2 (533 μS/cm) and Wouye P3N2 (670 gS/cm).

Sow (2001) shows that Lake Tanma has very high salinity levels. He shows the predominance of *Cyprideis*, which characterises a lagoon that is drying out in a semi-arid climate. *Cyprideis nigeriensis* is a species found in lagoons and estuaries where salinity reaches or exceeds 35^ for most of the year. This confirms the salinity of the water in Lake Tanma and the soil taken from the dried-up lake in 2018. This salinity is worsening with the irregularity of rainfall over the years, with repercussions for market gardening in the area.

Ndiaye (2008) talks about the lowering of the water table contributing to the rise of the saline

[10] **Céane**: a traditional well, often shallow, which can be tapped manually.

wedge, polluting several wells bordering Lake Tanma. These findings were confirmed by our analyses in 2018, which showed wells with saline water (1,500 to 7,330 µS/cm).
Elouard et al (1977) say that Lake Tanma remained in communication with the sea until recently. In 2019, with global warming and the advance of the sea, we can say that there is still a strong marine intrusion into the lake. In 2018, the highest salinity is noted in the waters (Senegal River: 45,000 µS/cm, Taré P3: 19,000 µS/cm, Darou Mboum P1: 18,200 µS/cm, Darou Mboum P2: 14,500 µS/cm, Wouye lake water: 12,000 µS/cm, Darou Mboum P3: 9,800 µS/cm, Tanma P3: 7,330 µS/cm, Taré P1: 3,000 µS/cm, Tanma P2: 2,200 µS/cm, Taré P2: 2,120 µS/cm, Wouye lake water: 1,920 µS/cm which have severe salinity levels.
In general, the water used for irrigation should have a low to medium degree of salinity, i.e. an EC of 600 to 1,700 µS/cm (lenntech, 1998). Only Tanma P1: 1,500 µS/cm has a reasonable salinity for irrigation. We can deduce from this that the salinity problems in the *Niayes* stem mainly from the water, due to global warming (advancing sea, lower rainfall, lower water table, etc.). There are also anthropogenic causes, such as excessive pumping of groundwater, which contributes to marine intrusion and the rise of salt to the surface. Market gardening requires huge quantities of water. To cultivate an area of one hectare, a permanent flow water of 80 m^3 per day is required, drawn directly from the groundwater table, which is currently being replenished very slowly (Diagne, 2008). The once very high water table is sinking under the repeated onslaughts of pumping. A lot of land has been abandoned because it is infertile, but the main problem is poor management of alkaline and saline soils, and poor use of saline water in plant irrigation. It is important to take soil types into account, in order to isolate the crops best suited to these soils. In the *Niayes*, salinisation is thought to be due to rainfall deficits (lowering of the water table) and marine intrusion. The phenomenon of marine invasion, which can extend several kilometres inland, is a major risk for coastal regions. Under certain conditions, salt water spreads inland and contaminates the groundwater close to the sea (Morsli et al., 2017). The invasion of freshwater by saltwater leads to soil degradation and salinisation as a result of irrigation with this water (Hamza, 2015).
2018 results showed high water salinity in the northern *Niayes* zone, especially in Gandiolais, with higher salinity than in the other study sites. At Darou Mboumbaye, the EC varies between 9,800 and 18,200 µS/cm. Diatta (2004), says that agriculture is threatened in Gandiolais because of the rise in salinity accentuated by the opening of the breach in 2003. In the Ndioukis, the increase in the rate of salinisation has led to a fall in yields and the abandonment of many gardens adjacent to the cordon vive. The results for 2010 and 2018 highlighted the negative impact of salinisation on market gardening, largely due to the opening of the gap. There has also been strong marine intrusion into the Gandiolais, encouraged by the breach. The results of analyses carried out on the river in 2018 confirm this situation, with an EC of 45,000 µS/cm.
Diallo (2005), says that analysis of the evolution of the Gandiolais water table has shown some very worrying results. The level of the freshwater table is falling year on year, with inestimable consequences for water resources and socio-economic activities. These changes are currently holding back local development in the Gandiolais coastal area. These results confirm our arguments about the drop in the water table in our study sites, contributing to the rise in salinity.
The results of Jacoutot (2006) in the Darou Mboumbaye sector show a fairly high salinity, with an average of 4,900 µS/cm (2.6 g/l) between Mboumbaye and Mbao, on the edge of the lagoon, which is permanently very salty. In the southern part of the sector, salinity reaches as

high as 9,520 µS/cm (5.40 g/l).
Salinisation has led to land abandonment (Lake Wouye 92 ha, Lake Tanma 4,720 m^2, Taré 8 ha, Darou Mboumbaye 2,250 ha), lower yields (Lake Wouye minus 1 tonne after 2008, Lake Tanma minus 1 tonne for more than 10 years, Taré minus 20 tonnes in 2017 and 2018, after the opening of the breach, Darou Mboumbaye minus 12 tonnes in 2017 and 2018, after the opening of the breach).
Studies carried out after 2010 have confirmed our results. Ndiaye et al (2012) state that the Niayes zone provides the bulk of Senegal's vegetable production, but its agricultural potential has been seriously reduced by land degradation due to the climatic hazards of recent decades and poor cultivation practices. The salinisation of the soil, which is more marked in the lowlands of the southern part of the country, can be explained by the lowering of the water table and the intrusion of sea water. Added to this are the significant influx of sea spray in areas close to the ocean, and the effects of the Diama dam and the relief canal on the Langue de Barbarie at Saint-Louis. According to the study, soil salinisation in the *Niayes* of Pikine could be largely attributed to a rainfall deficit, which is the cause of the lowering of the water table and the intrusion of seawater inland. This was confirmed by our 2018 results in Lake Wouye de Malika, which is located in the Niayes of Pikine, with a water EC of between 1,920 and 12,000 µS/cm, and in the village of Darou Mboumbaye in Saint-Louis (9,800 to 18,200 µS/cm).
In this context, Diallo et al (2014), say that several irrigated areas are subject to serious salinisation problems in the Niayes of Saint-Louis, resulting in soil degradation and a drop in productivity. Salinisation has a negative impact on the physical, chemical and biological properties of the soil, as well as on plant yields. They confirm our analysis results from 2010 and 2018 in Gandiol (in the village of Darou Mboumbaye). Analyses of soil and water samples taken in a number of Gandiol villages (Mboltime, Kalassane, Keur Saer, Ndoye Diagne, Rimbakh, Deggu Niaye and Ricotte) show that the pH varies in the wells from moderately acidic (Mboltime), neutral (Kalassane, Keur Saer, Ndoye Diagne, Rimbakh) to slightly alkaline (Deggu Niaye and Ricotte). The EC is saline at Keur Saer, very saline at Kalassane, Mboltime and Ndoye Diagne, and extremely saline at Deggu Niaye, Rimbakh and Ricotte with extremely high levels at the last three sites (between 4,414 and 5,843 µS/cm). Our 2018 samples show a change in water salinity in Gandiol, with an EC of 9,800 to 18,200 µS/cm. soil analysis results showed that there is a significant difference between the pH and EC values between water and soil. The pH is higher in the soil while the salt concentration is higher in the well water, just as our 2018 results showed, in contrast to 2010 and 2014 when there was an increase in pH (8.87 to 9.34 for soil, and 8.05 to 8.19 for water) and EC, (48 to 124 µS/cm for soil and 9,800 to 18,200 µS/cm for water). The study showed that water salinisation is widespread throughout the northern part of the *Niayes*. The less marked salinisation of the soil could be explained by the use of manure by producers to combat salinity.
According to the results obtained, the salinity has contaminated the water as much as the soil. The particulate structure associated with the sandy texture of the soils does not help matters insofar as irrigation water reaches the water table directly. Diallo et *al* (2015) produced some interesting results in their study of soil salinisation in the northern part of the Niayes. Their results showed neutral to slightly acid soils and very acid to slightly acid soils. This situation has changed in 2018, with alkaline soils in almost all of the *Niayes* area.
For electrical conductivity we have non-saline, slightly saline and very saline soils. Our 2018

results show non-saline soils in the north (48 to 124 µs/cm), with a higher EC in the village of Darou Mboumbaye, certainly due to the salinity level of the well water (9800 to 18200 µs/cm). Niang (2017), obtained interesting results in the study of soil and water salinisation in the northern *Niayes* zone. His results showed values indicative of highly mineralised water. In the village of Mboumbaye, which is in the same sector as Darou Mboumbaye, the results gave an EC of 15,270 µs/cm in the water, which is in the same range as our 2018 results in Darou Mboumbaye (9,800 to 18,200 µs/cm). The 2017 results showed a more or less salt-saturated water table, unsuitable for irrigation, which appears to be reaching its buffering margins, and a non-static nature of the EC of the quaternary water table. These data are confirmed by the 2018 analyses, which show that the water is highly saline, making it unsuitable for irrigation. Only a few crops are producing satisfactory yields, which means that we can assess the causes of salinisation in our study sites, to gain a better understanding of the phenomenon.

1.2. Causes of salinisation at the various sites

1.2.1. Natural causes

Several natural phenomena have contributed to the salinisation of the target sites. In the long term, they have led to the drying up of several watercourses, the lowering of the water table, marine intrusion, the salinisation and subsequent degradation of agricultural land. These phenomena include drought and rising sea levels.

♦*♦ Climate drought

Drought is the main cause of soil degradation in the Niayes. A number of studies have been carried out to demonstrate the main role of drought in soil degradation. The FAO's preliminary report on : The FAO's 2003 preliminary report on "the assessment of land degradation in Senegal" shows that the drought of the 1970s was in some ways the starting point for a long process of land degradation in Senegal, particularly in the *Niayes*. Research by Ndao (2012) has shown that the dry spells of the 1970s, 1980s and 1990s were marked by a drop in rainfall. During this period, the spatial and temporal evolution of the Niayes was characterised by a reduction in the surface area of wetlands, in particular by the disappearance of permanently flooded areas (drying up of lakes Mbeubeuss, Reubeut, Warouwaye, Wouye, Tanma, etc.), in favour of a spatial extension of soils with partial hydromorphy and partial hydromorphy with salinity. The study shows that such environmental changes are the result of a persistent drought, the intensity, duration and extent of which have been unique in the climatic records of the last fifty years. The rainfall deficits recorded in recent years vary in severity from year to year. Volumes of water have been considerably reduced and rainfall is more irregular in time and space.

Ndong (1995), shows that from 1970 onwards, there has been a chronic deficit in rainfall across the whole of . By decade, the period from 1961 to 1970 had little rainfall, with only two surplus years, 1967 and 1969. From 1961 to 1965, there were five years with little 104 rainfall, with 1968 and 1970 being the most deficient years of the decade. From the 1970s onwards, a period of drought began, characterised by its duration, intensity and extent. The most drought-prone years in the 1971-1993 period were 1972, 1973, 1976, 1977, 1983, 1984, 1990, 1991 and 1992. The 500 mm isohyet descends from 180 to 200 km south of the border with Gambia. The 800 and 1,000 mm isohyets descend by around 120 km further south compared with the 1961-1990 period, while the 1,200 mm isohyet only affects a very small portion of the territory, showing the migration of the isohyets southwards. Average annual rainfall varies between 200 mm in the north and 500 mm in the south, with an uneven distribution of monthly and daily rainfall over the year. Trends in inter-annual rainfall from 1930 to 2012 show that there is a high degree of climatic variability in the area, with periods of both surplus and deficit rainfall, particularly the long period of deficit rainfall in the Sahel since the 1970s (PADEN, 2017). This increases the risk of drought.

Aguiar (2009) points out that the Sahel has suffered years of drought in recent decades, particularly in the 1970s, 1980s and 1990s, when rainfall fell by around 40% (Hulme *et al.*, 2001). Between the periods 1950-1969 and 1970-1989, average annual rainfall fell by 46% in Dakar, 39% in Thiès, 43.7% in Louga and 35% in Saint-Louis, giving a regional variation of more than 41%. The average groundwater level has fallen by between 1 and 5 m, or between 7 and 30 cm/year, since 1965 (SGPRE, 2003). A decline in the level of the aquifer that is compromising market gardening in the Niayes, putting the region's food self-sufficiency objectives (which produces more than 80% of Senegal's vegetables) further out of reach (Tall and Fall, 2001). Evapotranspiration is the main factor in groundwater discharge. Years of drought have severely affected the Niayes' hydrographic networks. Several of these surface waters (Lac Tanma, Wouye, Mbeubeussé, etc.) dried up. Only a few (Wouye) regained their characteristics during the rainy seasons of 2000, 2002, etc. The drying up of surface water has been accompanied by a drop in the water table, causing a rise in the salinity level, accentuated by excessive pumping of the water table for irrigation, drinking water needs, etc. Drought as a result of global climate change is leading to a rise in general sea levels.

❖ **Effect of sea level rise on salinisation**

The melting of the glaciers at the poles and the expansion of the waters result in greater pressure from the marine waters on the continent and on the coasts. The pressure on the sandy facies (study sites) explains the greater spread/extension of the salt wedge on the loose sedimentary rock: the sand.

This is confirmed by research carried out by the Ministry of the Environment in 2006, which shows that coastal erosion affects almost all of Senegal's major coastal towns. The rates of shoreline retreat observed vary, but average between 1 and 2 m per year for sandy beaches (UNESCO, 2019). At Taré, in an area where the Langue de Barbarie is devoid of vegetation and therefore does not reduce the force of the north-westerly wind on the lagoon, cliff recession reaches 1 m/year (Sy et al, 2013).

These dynamics are strongly associated. The rise in sea level is the cause most directly linked to climate change (SSC, 2015). Based on records from the tide gauge in Dakar, it is estimated that over the last century, sea level has risen by an average of 1.4 mm per year. A second factor that could be climate-related is the occurrence of very strong swells, which are responsible for cuts in the coastal spits accompanied by very rapid retreat of certain parts of the coastline (Ministry of the Environment, 2006).

All these factors are at the root of the recurring problem of marine intrusion in our study sites (it should be remembered) and even throughout the *Niayes* area. Senegal is facing an invasion of its rivers, water tables and soil by marine water. Since the drought that set in at the end of the 1960s, river flows have fallen sharply (Ministry of the Environment, 2006).

Kane (2005) has shown that the average interannual flow of the Senegal River has continued to fall: 771 m^3/s in 1968, 765 m^3/s in 1971, 715 m^3/s in 1983 and 650 m^3/s at its mouth in 2019. This reduction in river flow, combined with the very low gradient in the coastal zone, has allowed seawater to rise up the river, favoured by the breach. The drought has led to a drop in the piezometric level and an advance in saline intrusion (salt wedge), with a strong presence of anions (Chloride, 13,595 mg/l) and cations (Sodium, 8,034.18 mg/l), which are the dominant ions.

Since 1968, the piezometric level has fallen by 5 to 10 m in the nappe on the north coast. In the sub-basaltic sands aquifer, between 1947 and 1987, the salt wedge advanced at an average rate of 35 m/year, leading to the abandonment of several boreholes (Gaye et al, 1989). The rainfall deficit recorded since the early 1970s accelerated the first processes leading to the salinisation and acidification of estuarine soils. This chemical degradation of the soil has destroyed or reduced any possibility of developing the land. The situation is exacerbated by the high demand on groundwater, particularly in the *Niayes* region. Water (Lake Wouye, chloride (232.40 to 610), sodium (92.05 to 2072.34) and soil (alkaline soil and saline water) are increasingly subject to saline intrusion. We are witnessing a marine invasion along the entire coastline, with erosion of around 1 m per year (Sy, 2017). This phenomenon has been accelerated by the disappearance of several filaos trees along the coast (e.g. Lake Wouye), which helped to fix the sand dunes. Estimates show a rise in sea level of 1 to 3 m by 2100 (Ray, 2015).

Figure 52: Sea level rise from 2017 to 2100

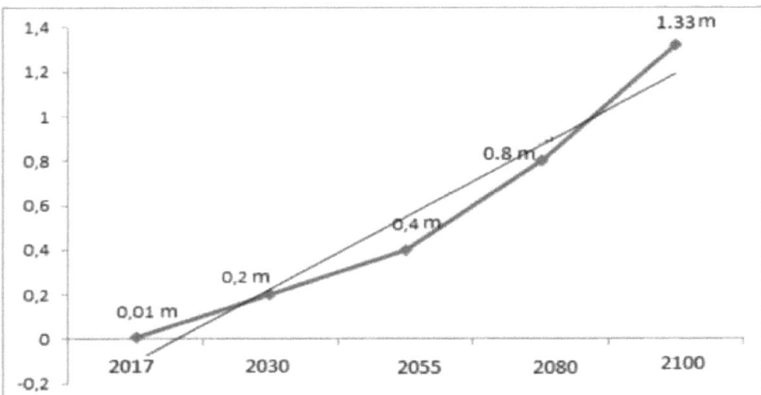

Source: M. Fall, Projection based on 2017 World Bank data

Apart from natural causes, the four study sites are faced with anthropogenic causes that impact soil and water quality: use of chemicals, excessive pumping of groundwater, opening of the Langue de Barbarie breach.

1.2.2. Anthropogenic causes

The anthropogenic causes here are the increasingly massive use of chemical products, excessive pumping of the water table and the opening of the gap in the Langue de Barbarie.

- **Use of chemicals**

Market gardeners in the *Niayes* are heavy users of chemicals (e.g. NPK). Many use them excessively, in an attempt to increase production and earn more income. Some use them to repair degraded or impoverished soils, which has consequences for the land and for market gardening in the area. Poor farming practices, such as the use of weedkillers, pesticides and soluble fertilisers, deplete and destructure the soil (the humus disappears).

The result is stratification, erosion and compaction of the soil, and its progressive inability to absorb and retain water. Soil destructuring leads to plant degeneration. Groundwater is affected by nitrate pollution, particularly in the urban area of Pikine in Dakar, where concentrations can reach 200 to 400 mg/l (Tandia, 2000). The main pesticides currently used in the Niayes belong to a few major chemical families:

- Organophosphates are synthetic compounds that degrade fairly quickly in the environment but have neurotoxic effects on vertebrates;
- Organochlorines (chlorinated hydrocarbons), such as DDT, are very chemically stable pesticides. DDT was used throughout the world to control insects, until it was discovered that it was not easily degradable and could concentrate in organisms at the end of the food chain, through bioaccumulation, with risks to human health. Its use banned in many temperate countries, but many are still found in aquatic environments. They are also still used in some tropical countries;
- carbamates, which are highly toxic, are used as insecticides and fungicides;
- pyrethroids, which are highly toxic to extremely toxic to fish and wildlife

aquatic. These are fast-acting products (shock effect after a few minutes), with varying levels of persistence and a broad spectrum (Bauer, 2007).

Table 4: Some pesticides reported by market gardeners

Trade name	Active ingredient (MA)	Concentration of MA	Formulation	Targets
		Organophosphates		
Malathion	Malathion	500 g/l	EC	IA
Tamaron	Metamidophos	600 g/l	EC	IA
Dimethoath	Dimethoath	350 g/l	EC	A
Metaphos	Metamidophos	600 g/l	EC	IA
Carbamates				
Lannate	methomyl	90 %	WP	I
Pyrethrino'ídes				
Decided	deltamethrin	25 g/l	EC	I
Symbush	cypermethrin	50 g/l	EC	I
Cypermethrin	cypermethrin	50 g/l	EC	I

Source: IDRC (2001)

Note: EC = emulsifying concentrate, WP = wettable powder, A = acaricide, I = insecticide.

IDRC research (2001) shows that chemical fertilisers with a formula of 10-10-20 (NPK) are commonly used, at doses varying between 150 and 600 kilograms per hectare depending on the crop. There is a wide range of organic fertilisers, but the problem of quality remains. In the Niayes, crop residues are mainly used to feed livestock, which creates a serious problem of biomass availability. Mineral fertilisation is used extensively in the area compared with the rest of the country, such as the groundnut basin. Sources of supply are fairly important in the Niayes agrosystem, even if access to these inputs remains difficult for some because of isolation and sometimes insufficient income. However, the products marketed by street vendors and even wholesalers are of dubious quality. This is due to the multiple repackaging and homemade manufacture of these agro-pharmaceutical compounds. This has an impact on soil quality.

Chemical fertilisers are widely used in market gardening to compensate for mineral deficiencies in plants. A wide variety of fertilisers are used, but not all of them are recommended, especially organochlorine pesticides. They are characterised by their toxicity, their persistence in the environment, their tendency to bioaccumulate in food chains, their ability to travel thousands of kilometres from their point of origin and their consequences for human health: neurological disorders, cancers, abnormal reproductive organs, mental deficiency in children, sterility. The most commonly used fertilisers are urea, 10-10-20, 10-10-15, 06-20-10, potassium sulphate and foliar fertilisers (IDRC, 2001).

According to Jacoutot (2006), fertilisation techniques are sometimes poorly mastered, especially when it comes to the use of chemical fertilisers by small-scale producers. They are widely used in market gardening to compensate for mineral deficiencies in plants. The quality and quantity of mineral fertilisers used vary according to the soil type of the substrate, the size of the farm and the type of crop. The most commonly used fertilisers are urea, potassium sulphate, foliar fertilisers, 10A, 15A, 10-10-20, 18-46-00, 06-20-10, 10-10-15, etc. 10-10-20, commonly referred to as "market garden fertiliser", is the most widely used in the area, as it is best suited to the needs of market garden crops, in doses varying between 150 and 600 kilograms per hectare depending on the crop and the budget of the market gardeners. Pesticides and fungicides are also used extensively to combat insect pests, but just like chemical fertilisers, these products have consequences for the soil and plants if they are not used sensibly.

Photo 20: Chemical products used by market gardeners

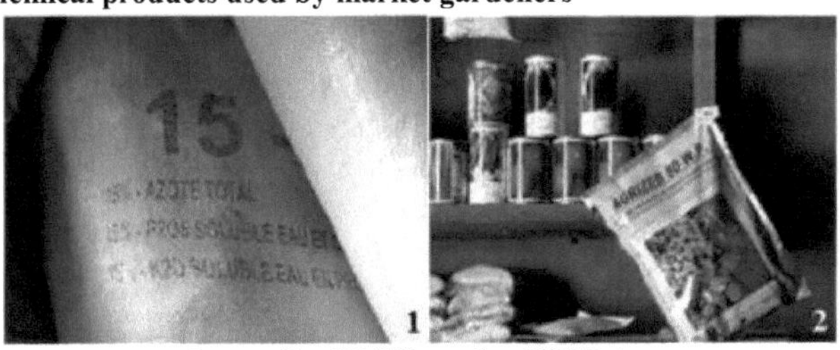

Source: Ray Weil December 2009 Potou

Image 1 shows N-P-K fertilisers, while image 2 shows boxes of seeds and plant packets of fungicide.

Despite the reduction in arable land, the use of high-dose chemical inputs seems to be an unstoppable option for intensifying production. There is a wide range of organic fertilisers, but the problem is one of quality. Livestock excrement is used extensively, particularly goat dung, which is spread on onion and carrot plants. Crop residues are mainly used to feed livestock, which creates a serious problem of biomass availability. Too much crop residue is lost to burning, burial or wind-borne dispersal in the fields. The study shows that the use of phytosanitary products is largely dominated by organophosphorus products, most of which are chemically related and highly toxic. Their high persistence means that they accumulate in food. Lack of information is responsible for the overuse of pesticides.

- **Excessive pumping of the water table**

Excessive pumping of groundwater causes saltwater to rise to the surface by thermocapillary action, through irrigation. Aguiar (2009) states that although natural factors (lower rainfall and evapotranspiration) are the main factors in removal of groundwater in 1974, the fall in the piezometric level has been accentuated by anthropogenic withdrawals, especially over the last 2 decades. The measurements he has taken in the wells show that daily fluctuations can vary from 0.50 m to 2 m. He shows that the sharpest piezometric falls were recorded during the drought of the 1970s and 1980s, with an average gradual decrease of almost 0.51 m every ten years. The projection made in relation to the Aguiar result reveals a piezometric drop of around 2.55 m from 1970 to 2020 (50 years). The drop in piezometry favours salinisation of the water table, with water EC varying in 2018 from 1920 to 12,000 pSZ cm at Lake Wouye, 1,500 to 7,330 µSZcm at Lake Tanma, 2,120 to 19,000 µSZcm at Taré, 9,800 to 18,200 µSZcm at Darou Mboumbaye.

The results showed that pumping water contributes to raising salinity levels, with high salinity in wells still in use (Wouye céane 1,920 µSZcm, Tanma P3 7,330 µSZcm, Taré P1 3,000 µSZcm, Darou Mboum P3 9,800 µSZcm). The high salinity of irrigation water contributes to soil degradation, leading to lower yields.

Douaoui and Hartani (2007) show that the use of saline groundwater for irrigation contributes to the degradation of soil quality, manifested by increased salinisation or sodisation. The proximity of the sea also has an influence on soil salinity through chloride-laden rainfall.

Ndiaye (2012), says that the phenomenon of soil salinisation is more marked in low-lying areas where the water table is shallow. The phenomenon of marine invasion, which can extend several kilometres inland, is a major risk for coastal regions that rely on groundwater for their water supply. Under certain conditions, salt water spreads inland and contaminates the groundwater close to the sea. This invasion of fresh water by salt water results in soil degradation and salinisation.

- **Opening of the breach on the Langue de Barbarie in 2003**

In October 2003, part of the city of Saint-Louis was threatened by a flood that reached a height of 175 cm. In response, an emergency solution was found: to open a relief channel (breach) on the sandy coastline of the Langue de Barbarie.

Photo 21: Overflow of the river at Saint-Louis

Source: Saint-Louis regional hydraulics department (Camara, 2008)

Photo 21 shows the river overflowing its banks in September 2003 before the breach was opened.

It was against this backdrop that the local authorities and the Director of the regional hydraulics department decided open an emergency "relief canal", on the night of 3 to 4 October 2003, 1.5 m deep, 4 m wide and 100 m long, dug around 7 km south of the city, with the aim of lowering the water level and thus preventing the flooding of the urban area of Saint-Louis, by facilitating the more rapid evacuation of the river's waters towards the sea. It was expected to reduce flooded areas by around 40%, particularly in the areas around Saint-Louis (Diatta, 2004).

Photo 22: Opening of the "relief canal" breach

Source: Saint-Louis Regional Water Department (Camara, 2008)

Photos 22 (1 and 2) show the first moments of the opening of the breach on the Langue de Barbarie (LB), to solve flooding problems in Saint-Louis.

The canal has lowered the water level by 1 m, preventing flooding (Diouf 2013). Opened with the aim of facilitating the flow water towards the sea during periods of high water and protecting Saint-Louis from flooding, the breach may have been a relief for the city, but not for the entire population of the area. It has caused a great deal of damage in certain parts of the area, and has profoundly altered the morphodynamics of the spit and the hydrological functioning of the Senegal River in relation to the influence of the sea. A great deal of research has been carried out since its opening, to show the consequences of such a decision taken in haste to resolve a situation, in the immediate environment of the region.

Camara researched the opening of the breach and its consequences in 2008. According to Camara, the breach in the Langue de Barbarie has had a huge impact on an environment where various activities are carried out side by side: fishing, tourism and agriculture. All these activities will be disrupted by the opening of the gap, changing the way they are organised in the Langue de Barbarie. Since the opening, the gap has widened considerably, some 7,000 m in 2020. This phenomenon can be explained by the effect of the north-south drift current. This current is becoming increasingly strong due to the intensity of the swell and the force of the river's flooding. This leads to changes in the hydrosedimentary processes, i.e. erosion of the seabed at the end of the "relief channel".

From a width of 4 metres when it was first opened, the canal reached a width of 250 m 3 to 4 days after digging (Camara, 2008). The width of the canal then increased to 750 m in 2004 and around 5,900 m in 2016, then to over 7,000 m in 2020.

Sy et al (2013) say that the rift will move further and further south, bringing with it the risk of villages and historic and biological sites disappearing, not to mention the threat to agriculture from salinisation.

Niang (2017), informs that at present, the speed of retreat has decreased in the southern part, which is logically explained by the enlargement of the channel and the dissipation of wave force.

Figure 53: Increase in the breach

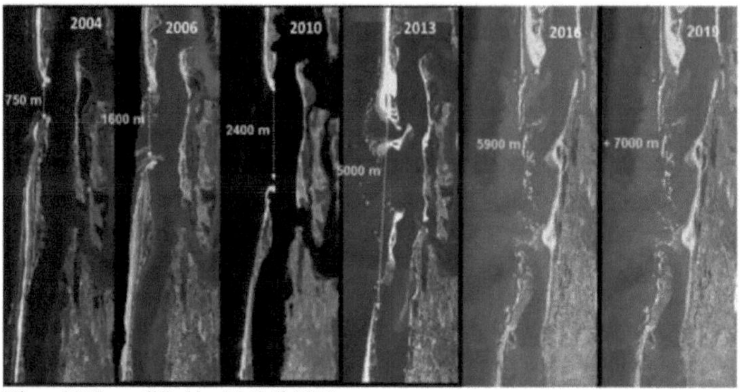

Source: Niang 2017, modified by M. Fall 2019

The image shows the evolution of the relief canal from 2004 to 2019-2020, which has grown steadily over the years, from 4 m in 2003 (when it opened) to more than 7,000 m in 2019-2020.

Camara (2008), considers the salinisation of the water table to be worrying. Since the opening of the breach, there has been significant salinisation of the water table, confirmed by our 2018 results at Darou Mboumbaye (9,800 to 18,200 µS/cm), and the people of the estuary are suffering as a result. Salinity in the river begins at the end of the flood and increases as soon as the Diama dam is closed. Our results show high salinity in the river in 2018, with 45,000 µS/cm. The salinisation caused by the Diama dam was accentuated by the opening of the breach. This salinisation is often due to the influence of the salt wedge, evaporation and marine submersion. The salinisation of the water table has led to a decline in market gardening, *in addition* to a reduction of 2,250 ha in the area under cultivation in Darou Mboumbaye, and in yields (less than 12 tonnes in 2017-2018, after the opening of the breach). Ndao (2012), shows that, from an agricultural point of view, market gardening is threatened in the Gandiolais region because of the rise in salinity accentuated by the opening of the breach. The study states that the breach will disrupt all the ecosystems in the area. The freshwater table in the area is becoming more saline, posing a serious water supply problem for local populations. In this sector, marine intrusion is threatening freshwater as well as certain agroforestry activities (market gardening). The severe drought of the 1970s, combined with the development of the river and the elimination of natural flooding, helped to reduce the level of the freshwater table. Since the opening of the breach, the freshwater table has been under threat. Exchanges between the lateral water tables and the river have been altered, resulting in salt contamination of the upper part of the river. Ndao adds that in Ndiouki-type market gardens (shallow depressions in low-lying areas with hydromorphic soils between the cordon vif and the semi-fixed yellow dunes), the increase in salt levels has led to a drop in yields and the abandonment of many gardens adjacent to the cordon vif. The well water associated with these plots has the lowest salinity (maximum: 1.8g/l, during a measurement campaign in February 2006). On plots near the river or at low altitude, however, the salinity is 2.9g/l. As a result, yields are lower and onion shoots are more or less yellow. Other plots, even closer to the river, have been completely abandoned for one or two years (well water at 12.5g/l salt).

The results from Ndao confirm our findings on the evolution of salinisation in the Gandiolais from the river inland, with a maximum of 9.1 g/l at Darou Mboum P1, the abandoned well closest to the village of Darou Mboumbaye.

Sy et al (2013) report that the saline lenticular sheet affecting the market gardening area is itself influenced by the southward progression of the breccia.

Crops in the Gandiolais are thus on hold, depending on their distance from this old branch of the river or their more or less perched nature. The economy of Gandiol, based especially on market gardening, is very much affected by this change. The current breach continues to function as a river mouth. This trend will disrupt market gardening in the Gandiol region, leading to excessive salinisation.

The results of Ndao (2012) confirm our findings in the village of Darou Mboumbaye, in Gandiol, with regard to salinisation of well water, lower yields and abandonment of plots.

Dieng (2010) speaks of the abandonment of almost all farms in the Gandiol region, due to the high salinity of the land.

This dynamic explains the exodus of the population other localities, departments, regions, neighbouring or distant countries (Spain for many young people, etc.), faced with the drop in income. The opening of the breach caused enormous difficulties. Resolving the flooding caused other problems, because the intervention was miscalculated. Now that the causes of salinisation have been determined, it is important to look at their impact on market gardening.

1.3. Consequences of salinisation on market gardening

Salinisation has a number of consequences for market gardening. The main consequences have been a reduction in cultivable land and yields. Soil alkalinisation and water salinisation have almost similar consequences. In all the selected sites in the Niayes of Dakar, Thiès, Louga and Saint Louis, we found a high rate of land abandonment. The results showed that almost all the land had been abandoned because of alkalinisation and salinisation. Many market gardeners have lost a large part of their land (Lake Wouye 60%, Lake Tanma 50%, Taré 40%, Darou Mboumbaye 80%), and some have even lost everything. In the case of Dakar, in addition to having lost land to salinity and alkalinity, the transformation of Lake Wouye into a retention basin has led to the flooding of almost all the plots of land near the lake, causing a drop in production and yields in all the selected areas: Lake Wouye minus 1 tonne after 2008, Lake Tanma minus 1 tonne for more than 10 years, Taré minus 20 tonnes in 2017-2018 after the breach, Darou Mboumbaye minus 12 tonnes in 2017-2018 after the breach.

1.3.1. Soil and water productivity

Salinisation on our sites leads to soil and water degradation, which limits market gardening. In recent years, we have seen a decline in soil productivity, with soils becoming increasingly alkaline, with pH values ranging from 7.2 to 9.34 in 2018, compared with 5.18 to 8.95 in 2010. Water is increasingly basic, with high salinity.

Since 2008, the situation of Lake Wouye has changed, with its transformation into a retention basin to solve the local flooding problems. Almost all the plots of land around the lake have been flooded (around 95% of them). As with the decision to open a breach at Saint-Louis, this decision by the government has had harmful consequences for the environment. Soil productivity has largely declined, with an increase in EC in 2018, which varies from 32 to 9630 µS/cm (in the four sites), compared with 2010 (1 to 192.7 µS/cm). The well water used to irrigate the market gardens had a high salinity level in 2018, lowering its productivity (1,500 to 19,000 µS/cm) compared with 2010 (1 to 10,920 µS/cm). The villages of Taré and

Darou Mboumbaye are among the victims of the opening of the breach in Saint-Louis, with high salinity of irrigation water in 2018 (2,120 to 19,000 μS/cm) in Taré and (9,800 to 18,200 μS/cm) in Darou Mboumbaye.

Taré's proximity to the old river mouth has always been considered an asset, with the river less than 30 m from the village, making it easier to grow vegetables in the area. In recent years, however, this advantage has become a handicap, with the opening of the breach in the Langue de Barbarie in 2003. As a result, salt is gradually encroaching on market gardening areas. This has led to a decline in soil and water productivity.

The opening of the breach only worsened the environmental problems that already existed in the two villages. The consequences of the drought of the 1970s were also felt, with a drop in the water table. In Taré and Darou Mboumbaye, the deterioration in climatic conditions has led to a drop in the water table and the deterioration and infertility of farmland. This salinisation and the lowering of the water table are limiting market gardening in the study sites. In the Niayes, several irrigated areas are suffering from serious salinisation problems, resulting in soil degradation ( ) and reduced productivity.

Figure 54: Villages of Taré and Darou Mboumbaye in relation to the breach

Source: Sy 2013, modified by M. Fall 2019

Diallo et al (2015) show that soil salinity reduces crop yields and that farmers sometimes stop working their land because of soil or water salinity. The study carried out soil analyses, which show that the pH varies from site to site, but there is no significant difference between the surface (0 - 20 cm) and the depth (20 - 40 cm). It is moderately acidic, neutral and alkaline. Salinity, on the other hand, varies according to site and depth. It is higher at the surface than at depth. It goes from very saline at the surface to saline at depth in some sites, with both horizons showing very high salinity. Similarly, saline soils at the surface and at depth in some areas, or from saline at the surface to neutral at depth. It can be said that at the sites studied in this study, the soils are very salty at the surface in most cases.

The study concludes that the soil analysis results showed that there is a significant difference between the pH and EC values between water and soil. The pH is higher in soil, while the concentration of salts is higher in well water. These higher concentrations of soluble salts in water may be linked to the sampling period. The water and soil samples were taken in May at the end of the dry season. In the dry season, overexploitation of water resources for crop irrigation is one of the main causes of salinisation. This leads to contamination of the groundwater close to the sea by the rising salinity, accentuated by the presence of the breach. This invasion of fresh water by salt water leads to soil degradation, salinisation and a drop in productivity.

The less marked salinisation of soils could explained by the use of manure by producers to combat salinity (Diallo et al, 2015).

The addition of nutrients to natural saline water reduces effect of salinity. This is one of the arguments used by market gardeners to justify the use of organic matter in irrigated plots (Snoussi, 1998). The high concentration of soluble salts in the soil can prevent seed germination and damage established plants, as well as impairing water uptake by plants.

Compared with the results of Diallo et al. (2015), which gave saline soils, our 2018 results show non-saline soils at all four sites, except for the soils taken from the surface of the dried Tanma Lake, which are highly saline, at 9,630 µS/cm. This situation can be explained by the locations where the samples were taken. The areas closest to the river can yield saline soils. Our study chose the plots that were still being farmed and the abandoned areas closest to the cultivated fields, justifying the lower EC of the soils (compared with their results), which are not yet saline but alkaline. Our results also showed a change in the EC of the four sites, which increased compared with 2010. This increase shows that the situation is worsening and that we are moving towards saline soils.

1.3.2. pH level and market gardening

At the four sites studied, the soil and water pHs were alkaline in 2018, with the soil (Lake Wouye 7.2 to 9.26, Lake Tanma 8.23 to 9.32, Taré 8.62 to 8.96, Darou Mboumbaye 8.87 to 9.34) and water (Lake Wouye 7.52 to 8.28, Lake Tanma 7.63 to 8.21, Taré 7.25 to 7.78, Darou Mboumbaye 8.05 to 8.19) alkaline. Compared with 2010, the results show an increase in pH, which varies from 5.18 to 8.95 for the soil and 6.97 to 8.32 for the water, due to marine intrusion, the use of chemicals and irrigation with saline water.

Arvalis (2014), says that the increase in pH following the application of a basic soil improver reduces the availability of trace elements such as boron, manganese, zinc and, to a lesser extent, copper. Phosphorus availability in the soil is also influenced by the pH . It is low at acidic pH levels (< 5) and basic pH levels (> 7). For several years now, market gardeners have been using chemical products to improve the quality of their soils, which are no longer producing as they used to. The results of analyses have shown the presence nitrate ions in the water, confirming the heavy use chemical products, with nitrate compounds ($NO3^-$).

Nitrate is a compound combining an anion with one or more cations. It is a salt found in chemical fertilisers, in the form of sodium nitrate, potassium nitrate, etc. Nitrate is part of the nitrogen cycle, particularly soluble in water and responsible for water pollution. It is dangerous because of its ability to transform into nitrite, which has recognised toxic effects. The presence of excess nitrates dissolved in water is an indication of pollution of agricultural origin (chemical fertilisers). In many countries, water intended for human consumption must comply with limit values (e.g. 50 mg/l in France and Europe) in order to be considered drinkable. The WHO recommends that this 50 mg/l threshold should not be exceeded. Nitrates are powerful eutrophying agents, and as such are considered to be environmental pollutants above the levels normally found in nature, which vary according to the type of natural habitat (Lamiot, 2014). The results of analyses at the study sites show, on the whole, quantities of nitrates in excess of the standard set by the WHO (50 mg/l).

In 2018, the results show heavy pollution of the water table by chemicals, with significant levels of nitrates in the water at the four study sites (Lake Wouye 50.44 mg/l in total, Lake Tanma 164.46 mg/l with 121.6 mg/l at Tanma P2 "abandoned well in the cultivated field", Taré 392.85 mg/l varying between 81.88 and 200.93 mg/l, Darou Mboumbaye 389.32 mg/l varying between 69.43 and 247.96 mg/l). The results show a higher level of water pollution in the northern Niayes zone (Taré and Darou Mboumbaye), which exceeds WHO standards (50 mg/l), with a higher level in Darou Mboumbaye (explaining the 80% of abandoned land). When an environment is not naturally buffered against acids (acid neutralisation), as in our case, the nitrate ion increases the pH of the environment and increasing acidification is inevitable (unless countermeasures are taken). Nitrates have become the new and worrying factor in the acidification of fresh waters, which has already led to significant "biotic impoverishment". It sometimes reaches or exceeds toxicity thresholds that limit the "growth, reproduction or survival" of individuals or species (Lamiot, 2014). Table 5 illustrates the recommended pH level for some crops adapted to neutral, acidic and basic soils.

Table 5: Recommended pH levels for some crops

Acid to Neutral 5 < pH< 7.5	Recommended pH	Neutral pH: 6.1 to 7.5	Recommended pH	Neutral to Basic 7.5 < pH< 9	Recommended pH
Aubergine	5 à 7	Pumpkin	6 à 6.5	Mint	7.5 à 8
Beans	5 à 7	Courgette	6.3 à 6.5	Garlic	
Parsnips	5.6 à 6.3	Turnip	6 à 6.5	Beetroot	
Cauliflower	5.5 à 6.5	Celery	6 à 6.5	Carrots	
Peas	5 à 7	Spinach	6 à 6.5	Cauliflower	
Soya	5.5 à 6.5	Melon	6 à 6.5	Beans	
Tomato	5.5 à 6.5	Chilli	6 à 6.5	Lettuce	
Sweet potato	5 à 6.8	Radish	6 à 6.5	Radish	
Potato	5 à 6.5	Tomatoes	6 à 6.5	Peas	
Cucumber	5 à 6.5			Tomatoes (if not too chalky)	
Peppers	5 à 6.5				
Parsley	5 à 7				
Asparagus	5.5 à 6.8				
Carrot	5.5 à 6.5				
Shallot	5.5 à 6.5				
Sorrel	5.5 à 6.5				
But	5.5 à 6.5				
Garlic	5.6 à 6.5				

Beetroot	5.6 à 7				
Cabbage	5.6 à 7.2				
Lettuce	5.6 à 6.5				
Onion	5.6 à 6.5				
Leek	5.6 à 6.5				
Broccoli	5.6 à 6.5				
Brussels sprouts	5.6 à 6.5				

Source: Mark (2015) and Fail (2019)

The results show that most of the abandoned land does not have recurring problems of salinity, but rather alkalinity. The results also show that vegetable tolerance levels hardly reach a pH above 7.5. This explains the deterioration of crops and the abandonment of certain plots by market gardeners, who believe that this is due to the salinity of the soil. The salinity is especially present in irrigation water and in lakes Wouye and Tanma. This situation is due to marine intrusion, which is becoming increasingly significant as the sea advances over the years. We can deduce from this that, in addition to the alkalinity of the soil, there is a contribution of salt to the plantations from irrigation water. The plants will have to adapt to the alkalinity, but also to the salinity. The fact that market gardeners are abandoning their plots shows that they are not succeeding. With the salinity of irrigation water, it's clear that in the long term, we'll have a serious problem with soil salinity.

The results of Diallo (2015) confirm this, with the presence of saline soils in Gandiolais. The change in EC at our target sites between 2010 and 2018 is also a reference point for the possibility of long-term soil salinisation. In the absence of a permanent water table, sandy soils such as those at Taré and Darou Mboumbaye are unlikely to become salty through irrigation (they do not retain water), but the high evapotranspiration in the area poses a problem (favouring salinisation). Medium-textured soils such as those at Lac Wouye and Lac Tanma retain water in their surface horizons, and this water tends to migrate upwards, attracted by capillary action, before evaporating and depositing its salts on the surface. The more salt-laden the water used, the more leaching water is required. Irrigated with salty water, they can be transformed into saline soils, each irrigation leaving on the soil the quantity of salts contained in the water used. These phenomena call for good drainage, which will nevertheless leave a certain amount of salts in the soil, which the plants will have to deal with. At the four study sites, the analyses showed an insignificant presence of salt in the soil in 2018 (Lake Wouye 290 to 670 µS/cm, Lake Tanma 49 to 9630 µS/cm, Taré 32 to 86 µS/cm, Darou Mboumbaye 48 to 124), with a higher amount of salts at Lakes Wouye and Tanma due to the texture of the soils, which retain water more than the sandy texture of Taré and Darou Mboumbaye, with high water infiltration. The results also show a 122% increase in EC at the four sites between 2010 and 2018, which could continue over time. Water always provides the soil with a certain amount of salts that the irrigated crop has to deal with (Durand, 1973). The risk of irrigating with salty water is that plants absorb water from the soil because it is soft, and there is salt in the root cells. The salt in the root cells attracts water to the plant.

The water moves along a gradient of water activity, from the compartment with the lowest osmotic potential to the one with the highest. Irrigation with saline water poses a threat to market gardening, causing a drop in yields. For this reason, it is important to assess the evolution of vegetable production.

2. Market garden production in the study sites

Market gardening is a profitable activity for farmers for many years, with production likely to exceed 30 tonnes/ha. It is important to highlight the changes in market garden production before and after the salinity problems, in order to assess the impact of salinisation on market gardening.

In the case of Lake Wouye, it is important to remember the situation prior to 2008 (when the lake was converted into a wastewater retention basin). In 2006, surveys showed that 35% of farmers earned between 50,000 and 100,000 CFA francs a year from market gardening, while 18% earned between 150,000 and 200,000 CFA francs, which means that market gardening is a profitable activity. But sometimes the crops are destroyed by parasites, pests or salinisation. Production was up to 3 tonnes per hectare in previous years, but the transformation of the lake into a retention basin in 2008 changed all that.

In 2010, some 93% of farmers lost their market gardens to the lake. The majority of market gardeners were unable to grow crops this year, which had an impact on their farm income. In 2018, as the lake receded, some farmers (2%) were able to recover their plots. Market gardeners say that the land closer to the lake is not productive because of the salt, and that many of them have cultivated at a loss.

With the rainfall in 2019, the land is still flooded (90.2% of plots). Market gardeners who are far from the immediate vicinity of the lake, in the northern part (7%), are still producing as before and earning their income from market gardening (from 150,000 to 200,000 CFA francs) per season. However, they have noted a drop in production compared with 2006, with less one tonne in 2017-2018 (500 kg/ha per month). Figure 56 shows market garden production at Lake Wouye before and after its conversion to a retention basin in 2008.

Figure 55: Market garden production at Lake Wouye before and after 2008

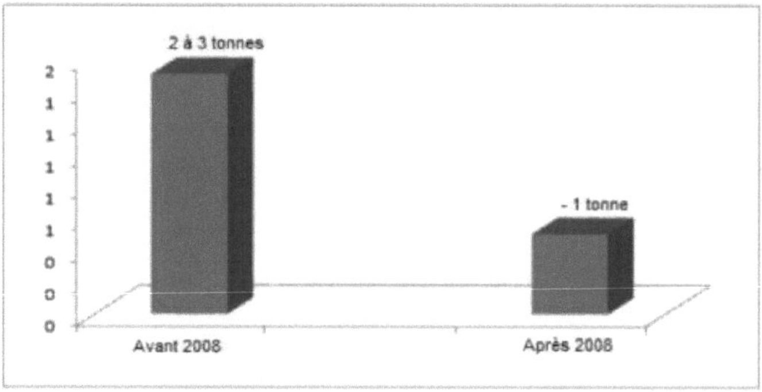

Lake Tanma and Lake Wouye have similar problems: salinisation of irrigation water and alkalinisation of market garden areas. This situation has repercussions on production. Results from Lake Tanma showed that before 1988 (when market gardeners began to notice the degradation of the land), production was 3 to 4 tonnes per harvest, with potato and sweet potato crops. After 1988, with salinisation, production fell to less than 1 tonne (around 900 kg).

Growers say that they no longer earn much money from market gardening, whereas before the degradation of the land, they could earn up to 500,000 CFA francs per season, enabling them to meet their needs and support their families. Some of them have even built their houses thanks to this activity, which was really profitable. Today, they say that they often cultivate at a loss, because the land is no longer as fertile as it used to be, and because of the diseases that attack the plants, such as iron chlorosis.

Figure 56: Market garden production at Lake Tanma before and after 1988

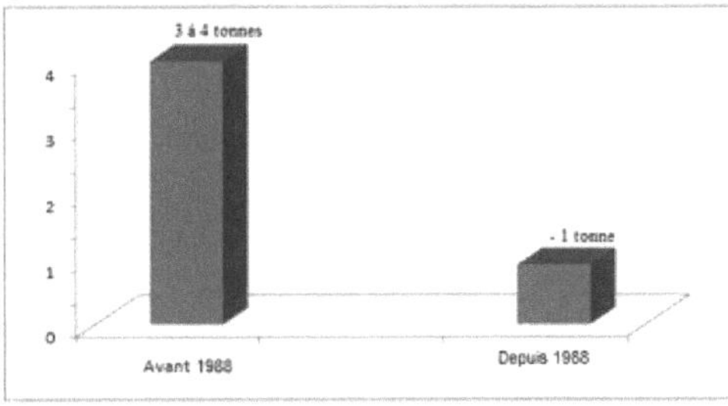

At Taré, before the breach was opened in 2003, production was 30 tonnes or more per hectare. In 2012-2013, production was 20 tonnes per hectare (with a deficit of 10 tonnes). Market gardeners say they observed a drop in yields per hectare after the opening of the breach. They say that it was only after 2003 that they noticed that the land was no longer producing as before. This situation has worsened in 2018, with production falling even further (to less than 20 tonnes). The drop in production is a great loss for them, because before, they could earn more than 500,000 CFA francs per harvest, but now it varies between 300,000 and 500,000 CFA francs, or even less.

Figure 57: Market garden production at Taré after the opening of the breach

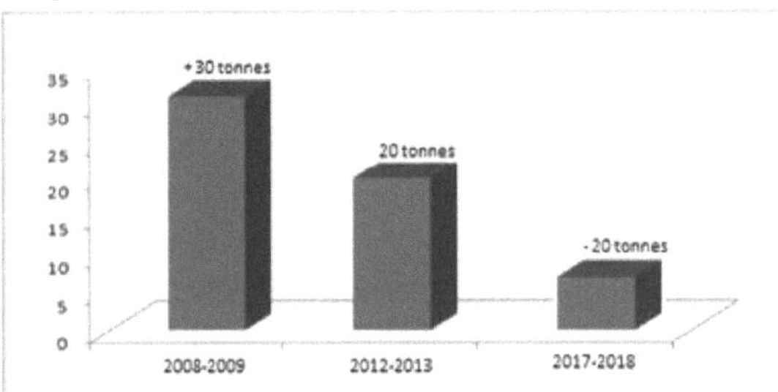

Like Taré, Darou Mboumbaye suffered the same damage caused by the opening of the breach in 2003. Before the breach, production was 30 to 40 tonnes per hectare, compared with 15 tonnes per hectare in 2013, and since the breach (2013), it has been 12 to 15 tonnes per hectare. The market gardeners say that the breach is a great detriment to the market gardening activities of Darou Mboumbaye, with a huge financial loss. They say they used to earn up to 500,000 CFA francs or more from market gardening, but now they earn much less (less than 500,000). In 2018, the situation is worse, with less than 12 tonnes per hectare. Figure 59 shows market garden production in Darou Mboumbaye after the Langue de Barbarie was breached in 2003.

Figure 58: Market garden production in Darou Mboumbaye after the breach in 2003

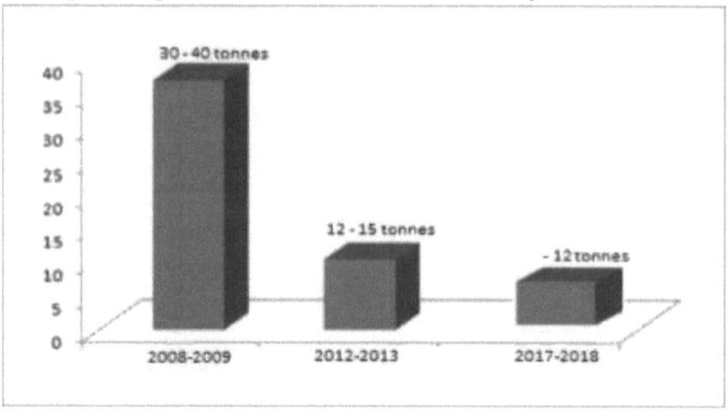

In 2018, the situation has not improved; projections show a worsening of the chemical situation in the face of the advancing sea (coastal erosion). Although market garden production is satisfactory at national level (macro), it is experiencing serious problems at micro level (traditional farms in the Niayes), with a sharp decline over the years. It is important for market gardeners to combat salinisation in the Niayes to boost production again. In 2019, individual strategies are being implemented by some market gardeners in the villages of Taré and Darou Mboumbaye to try and solve the problem. Others are using pumps to tap wells just a few metres from their market gardens, to get water that is more drinkable and therefore more suitable for irrigation.

3. Combating soil salinisation

Since 1960, major efforts have been made to revive and diversify agricultural production through the implementation of several agricultural programmes. Despite encouraging results, the initiatives taken have shown their limitations (LADA, 2003).

In 1997, Senegal's pedological office estimated that 1,700,000 ha were affected by salinisation in Senegal. Two types of action are being taken to combat this phenomenon: mechanical and biological/chemical. The first (mechanical) technique involves building anti-salt dykes and reinforcing the existing diking network to prevent the intrusion of sea water. This technique, used in the Saloum before 1960 and improved by research, has been applied in several projects (PRS, PIDAC, DERBAC, PROGES, etc.). Research has developed ridge-cultivation techniques, created early rice varieties and, to regulate water circulation, rack (ORSTOM, CIRAD, ISRA, etc.). In the biological field, as part of a partnership between ISRA and Ben Gourion University, halophilic species (Israeli varieties of *Tamarix sp*) were used to reforest the saline lands of the Sine Saloum. This programme has achieved convincing results in terms of saline land recovery.

In this context, phosphogypsum has been widely used to combat salinity in irrigated crops. NGOs have also contributed to the fight against the chemical degradation of land through actions to combat the degradation of plant formations, salinity, acidity and erosion.

In the early 2000s, the Senegalese government set up PAPIL with the help of partners such as the African Development Bank (AfDB) and the Islamic Development Bank (IDB). The regional branch of the small-scale local irrigation support project (PAPIL) recovered more than 2,210 hectares in 2012, out of a total area of 171,724 hectares (26.42%) of soil affected by salinity in the Fatick region (Diop, PAPIL, 2013). In 2015, PAPIL was replaced by the Multinational Programme to Strengthen Resilience to Food and Nutrition Insecurity in the Sahel. One of the aims of this programme is to reclaim thousands of additional hectares of

salinised land in the Fatick region by 2020. "The fight against salinated land requires a lot of resources and long-term programming" (Thiam, caraf 2017).
Improving the properties of saline soils and the salinity tolerance of plants would facilitate the restoration of saline lands, through a combination of agronomic and biological methods (Fall, 2016). In his research carried out in 2016 in Fatick, Fall gave some very interesting results on the recovery of saline lands in Senegal. The results show that the use of groundnut shell can be a good alternative, for a natural input, with the dose 6 T/ha that allowed the best growth of plants in saline conditions. The results showed that microbial inoculation (rhizobia alone, CMA alone or rhizobia + CMA), groundnut hulls (6 T/ha) and their combination improved the growth, nodulation and mycorrhization of plants grown under saline conditions in the greenhouse and in the field. However, there was no significant difference between plants amended with hulls alone and those inoculated and amended. However, the best growth and survival rates were obtained with the combination of microbial inoculation and groundnut hulls. The results suggest that *Senegalia senegal* (Syn. *Acacia senegal*) and *Vachellia seyal* (Syn. *Acacia seyal*) have the same salinity tolerance as *Prosopis juliflora* and that the combination of microbial inoculation (rhizobia + CMA) and groundnut hulls var 7333 (6 T/ha) increases their potential to restore salinity-affected soils in Senegal.
The results of Fall (2016) show that organic fertilisers are the best solution for combating soil salinisation. It would be important to continue research at our sites to monitor developments in the situation, but above all to conduct *in situ* experiments on techniques for adapting crops to alkaline soils and saline water. The experiments should focus on new irrigation techniques that use saline water adversely affecting plants. Drip irrigation could be an option, especially solar drip (kondenskompressor), which could be less costly. In the present case of our study, where the results show salinisation of the irrigation water and not the soil, the addition of organic matter could be a preventive rather than a curative method.

4. Prevention of salinisation

In the context of our study sites, it is above all necessary to find preventive methods to prevent the salinisation of alkaline soils and to find methods of recovering and conserving rainwater (even if rainfall is becoming increasingly rare). The other solution is to limit the evapotranspiration of irrigation water as much as possible (avoid open basins where well water is stored), and invest in closed storage basins. The other essential element is to grow species that are tolerant of alkaline soils, but above all saline waters, for better yields. As soon as irrigation is introduced, precautions must be taken to ensure the quality of the water supplied and the disposal of drainage water. Irrigation water must not contain more than 2 g/l of salts, or have a pH of less than 0.5. conductivity of more than 3 mmhos/cm. Their S.A.R. must not exceed 15 meq/l, and their residual alkalinity must not exceed 2.5 meq/l (Le Brusq, Loyer, 1983).
In our four study sites, 33% of samples exceeded 15 meq/l for SAR results (Wouye lake water 25.44 meq/l, Taré P3 25.46 meq/l (which is in an abandoned field), Senegal River 58.78 meq/l, Darou Mboum P1 28.61 meq/l (which is in the abandoned field closest to the village). The results show that, on the whole, the water from the wells sampled has a high S.A.R. and is not of good quality for irrigation (especially in the northern Niayes zone), which can cause serious problems for the soil. High sodium levels alter soil structure and permeability. The risks of salinisation and alkalinisation are not independent. The higher the salinity, the greater the risk of alkalinisation (Hubert, 1984-1998). Sandy soils tolerate moderately salty water

better than clay soils (high retention capacity), but the problem lies mainly with the plants. Many vegetables are not tolerant of saline water. A suitable irrigation system is therefore essential to avoid growing crops at a loss. In our case, the use of new irrigation systems (drip irrigation) could limit the effects of saline water on plants.

Table 6: Constraints and solutions to water salinisation and soil alkalinisation

N°	Stress s	Solutions
1	Salinisation of irrigation water	- creation of basins (preferably closed impluvium basins) to limit evapotranspiration - rainwater collection in basins (during rainfall)
2	soil alkalinisation	- reduce as far as possible, or stop completely, the use of chemical products - organic amendment (composting) - crop association
3	Decline in soil fertility	- Promote organic fertilisation, - use the furrow technique - Facilitate access to fertilisers (preferably organic), such as Azadirachta indica (Neem in Wolof), to combat insects. - regular long-term application of manure (to reclaim abandoned land) - adding sand to flooded land on Lake Wouye (to lighten the weight of the soil and recover areas that are not excessively flooded)

Soil improvers will be essential for recovering the areas lost at the four study sites. Long-term application of organic soil improvers and 129 mineral soil improvers, if necessary, will have to be carried out regularly over several seasons (years) to achieve satisfactory results. This will allow the soil to recover its nutrients and acidify the soil by lowering the pH level. The role of organic soil improvers is not to replace fertilisers, but to improve the physical resistance of plants and soils depleted of nutrients or weakened by the massive use of pesticides and agrochemicals. Organic soil improvers are plant-based. They lighten heavy soils, give body to light soils and replenish the soil's stock of organic matter and humus. Through their gradual mineralisation, they provide long-term nourishment for plants, without the risk of leaching, while ensuring better air and water circulation. They nourish the soil before nourishing the plant, once the organic matter has been broken down into assimilable mineral substances. The main organic soil improvers are livestock manure and compost. They can be used in all situations and are incorporated into the soil. Other materials can also be used: straw, bark, waste composts, green manures, commercial potting soils. We recommend using organic matter that has already decomposed. Organic matter that is too fresh uses the nitrogen in the soil to decompose, to the detriment of the plant. This leads to a nitrogen deficiency known as "nitrogen hunger" (SNHF, 2018).

In the case of our study sites, it would be necessary to use mineral amendments on areas that are still suitable for cultivation and those that have been lost. Mineral amendments improve certain physico-chemical properties of the soil, such as correcting pH or improving soil structure, while making it easier to work. They also enable plants to absorb nutrients more easily. The main mineral soil improvers are: lime, gypsum, wood ash, sulphur, iron sulphate, sand, clay, marl, etc. Lime can raise the pH of a soil that is too acidic, while iron sulphate can acidify a soil that is too basic. In the case of our study sites, the use of sulphur would be the most appropriate, to lower the pH and acidify the soil. In the case of flooded soil at Lake

Wouye, the flooded areas will need to be backfilled with sand to lighten the heaviness of the soil.

Conclusion

The *Niayes* are a special ecosystem made up of inter-dune basins located on outcropping to sub-outcropping groundwater all along the north coast from Dakar to Saint-Louis. They are located at the interface between coastal dune systems and the Ogolian erg, which explains why they belong to both coastal and continental ecosystems. This very fragile zone is currently affected by constraints linked to climate drying, human pressure, salinisation of the water and alkalinisation of the soil. This has a number of consequences:

- significant lowering of the water table;
- drying up of lakes Mbaouane, Tanma, Mboro, Mékhé ;
- soil and water degradation ;
- lower yields.

To sum up, trends in water resources show a worsening of the problems of resource availability, in terms of both quantity and quality. In addition, the high spatial and temporal variability of these water resources is becoming a major constraint on their use. The deterioration in climatic conditions and repeated droughts have had a number of consequences:

- fall in the piezometric level of certain water tables due to a shortage of supply and heavy evaporation;
- intensive exploitation of several aquifers to supply water to the city of Dakar and the surrounding area;
- threat of salt-water intrusion, especially along the coastal strip;
- rise of the salt tongue along the Senegal River, encouraging salt water intrusion into certain aquifers;
- deterioration in the chemical quality of water in the coastal strip (Niayes) following the drop in the piezometric level, which has led to an advance of the salt front.

The Niayes ecosystem is being disrupted by the degradation of soils weakened by intensive farming and the excessive lowering of the water table, exacerbated by the decline in rainfall. The loss of biodiversity is visible, with the disappearance of many plant and animal species. There has even been a deterioration in the strip of filao along the coastline, which is very important in preventing advance of the dunes.

In the 1970s, rainfall fell throughout the country and most of the lakes, ponds and marshes along the northern coast dried up. A wave of migrants ended up in the Niayes, where the water table is at or below the surface.

At present, the Niayes have no real river flows. However, their morphology suggests the existence ancient exoreic fluvial valleys perpendicular to the coast. There are numerous lakes in the southern part (Dakar, Thiès) that were occupied by the sea during the Nouakchottian transgression: Mbawane, Tanma, etc.

Many of these lakes have lost their size and importance. On the other hand, in the central and northern Niayes, there are only ponds whose duration in time and space remains largely dependent on rainfall.

The persistent deterioration in climatic conditions means that surface run-off is becoming increasingly scarce; water resources in the Niayes come mainly from the water table in the quaternary sands, which is of vital importance because of its multiple uses. It is used to supply water to the local population, particularly the city of Dakar, to feed animals and for the agricultural needs that make the region so important (Fall, 2001).

Crops such as onions, potatoes, carrots, turnips and cabbage are grown on the margins of these depressions. The gradual drying out of the wet depressions has led to the relocation of the market gardening areas from their margins to the slopes of the dunes. The plants are irrigated from "céanes", traditional wells varying in depth from 1 to 10 metres. The reduction and scarcity of rainfall are the fundamental factors that have triggered all the current natural and man-made processes. The environment is experiencing an unprecedented climatic deterioration in terms of its severity and duration. This has been accompanied by changes to the landscape.

It is important to note that all the regions covering the Niayes are facing the same problems in terms of rainfall deficits, degradation and burial of lowland soils, lowering of the water table, salinisation of water and alkalinisation of soils. The physical environment of the Niayes faces numerous problems, which have repercussions for all the regions around it.

Any solutions should take into account the general rather than the specific context. All these factors have repercussions on the socio-economic activities carried out in the area, particularly market gardening. Degradation of the land and groundwater, with the gradual advance of salt, falling production and yields, increasing poverty. Many constraints still need to be overcome (salinisation and alkalinisation of water and soil), despite the significant progress made thanks to the increased technical skills of farmers, in conjunction with the combined efforts of research, management and public and private sector companies.

A large proportion of vegetable production comes from family farms, particularly in the Niayes areas (centre and north). These are small farms where irrigation is the main occupation of the workforce, and where growers are often faced with problems of financing, input quality and production. It is in the face of this lack of resources that other environmental problems such as salinisation of water and alkalinisation of soil, are added to the difficulties faced by producers. The factors that lead to an excessive accumulation of salts in the soil can be natural or anthropogenic. The natural factors that influence water salinity and soil alkalinity are climate, soil parent material, vegetation cover, vegetation type and topography. Anthropogenic (human-induced) factors that can lead to salinisation or alkalinisation include irrigation with salt-rich water, the raising of the water table due to human activities (excessive pumping of the Quaternary sand table), the use chemical fertilisers and soil contamination by salt-rich water.

Inadequate irrigation practices (using irrigation water rich in salt, for example) and poor drainage lead to salinisation. Salinisation and alkalinisation are often associated with irrigated areas where low rainfall, high evapotranspiration or soil textural characteristics prevent salts from leaching out of the soil, accumulating in the surface layers. Irrigation with water with a high salt content exacerbates the problem. In the Niayes area, salinisation can be linked to the over-exploitation of groundwater, driven by increasing urbanisation, industry and agriculture. Over-extraction of groundwater can lower normal water tables and lead to seawater intrusion. The results showed an increase in production over the years, at the macro level, and a downward trend in production and income at the micro level, with repercussions for people's quality of life. This shows that water salinity and soil alkalinity have a negative impact on market gardening in the Niayes, confirming our hypothesis 2: "salinisation has led to a decline in market gardening activities and yields". For this reason, strategies have been put in place at national level to combat soil salinisation. These include the construction of anti-salt dykes, reforestation campaigns, and training and awareness-raising campaigns for the population on organic fertilisation and drainage, to recover degraded land. For our four study sites, we need strategies for recovering alkaline soil and finding a better irrigation system to limit the effects of salt water on plants.

General conclusion

Land degradation has been identified in all regions of Senegal, particularly in the *Niayes*. The phenomenon is manifested by a decline in soil productivity and heavy exploitation of agricultural land, while the population continues to grow at an uncontrollable rate, with some 1,138,956 inhabitants in 2002 (Camara, 2010). The combined effects of physical and human factors have led to an impressive decline in soil production potential and to an intensification of salinisation and alkalinisation phenomena.

For some years now, crop yields of less than 1 tonne/ha have been regularly recorded in the central Niayes area, more specifically in the Thiès region (at Lac Tanma), despite the almost total disappearance of fallow land. In a large part of the Niayes, changes in land use have been marked by the abandonment of land formerly given over to rain-fed crops and the conversion of local populations to market gardening. In any case, these changes have led to land degradation through salinisation (soil of the dried-up Tanma lake) and alkalinisation (Wouye, Taré and Darou Mboumbaye lakes).

The seriousness of water salinisation and soil alkalinisation is well known to local populations, who assess the state of soil degradation on the basis of its level of productivity and the damage caused to crops. Our four study sites were used as a reference point for the overall situation of salinisation of irrigation water and alkalinisation of soils in the Niayes, to show the evolution of the phenomenon in these environments and its consequences on market gardening activities.

The soils of Lake Wouye and Lake Tanma are gradually losing their natural value, with changes leading an accelerated deterioration and transformation of their former appearance, with repercussions for market gardening, with a drop in cultivable areas and yields. In the villages of Taré and Darou Mboumbaye, while the opening of the breach saved the island of Saint-Louis from flooding, it exacerbated the problems of water salinity and soil alkalinity in the market gardens of the two villages. The problems of water salinity are getting worse, with the breach getting bigger and bigger. This has not been helped by the general rise in sea levels, which is accentuating the di:usion of the salt-water wedge in the sandy coastal facies of the Grande Côte.

The methodology chosen for this study (direct observation, interviews, sampling, analysis, etc.) was extremely useful. It enabled us to identify the salinisation of water and the alkalinisation of soil, the natural and anthropogenic causes and the consequences for market gardening activities. The work undertaken as part of this research has produced the following results:

- the government's decision to turn Lake Wouye into a reservoir has destroyed much of the market garden;
- rising water levels in Lake Wouye have reduced the area available for cultivation;
- the market gardens around Lake Tanma contribute to its degradation, with the use of chemical products;
- the use of chemicals has contributed to soil degradation and water

pollution (lakes, wells, etc.);
- Climate warming has caused ocean water levels to rise, increasing the speed at which the salt-water layer extends into the lenticular nappe of the north coast, where the lowlands (Niayes), the subject of this research, are located;
- Coastal erosion has also accentuated marine intrusion;
- marine intrusion has led to salinisation of the water and alkalinisation of the soil;
- salinisation of water and alkalinisation of soil have led to a reduction in the area available for cultivation;
- the shrinking of arable land has led to a fall in agricultural yields;
- the fall in yields has led to a fall in market gardeners' incomes; - the fall in market gardeners' incomes has led to the abandonment of market garden areas affected by the salinity of the irrigation water and the alkalinity of the soil;
- The opening of the gap in the Langue de Barbarie has increased the salinisation of the water and the alkalinisation of the soil in the villages of Taré and Darou Mboumbaye;

The authorities and the local population urgently need to become more aware of the phenomenon of soil degradation. Malika, Diender, Taré and Darou Mboumbaye have lost much of their arable land. Being in the Sahel, the authorities and the population must not contribute to the degradation of natural resources (soil, lakes, groundwater, etc.) through their decisions and behaviour. It is important to find ways of preserving natural resources (soil, water, etc.), which are the victims multiple aggressions (natural and man-made), if market gardening in the Niayes is to continue. If the loss of arable land continues, market gardening will disappear. At Lake Wouye and Tanma, farms have all but disappeared, due to both natural (drought, salinisation, alkalinisation, coastal erosion, etc.) and man-made (chemicals, bad policy decisions, disappearance of fallow land, etc.) phenomena. If nothing is done about the salinisation of water and the alkalinisation of soil, market gardening is in danger of disappearing in the Niayes, where marine intrusion is very strong.

This threat is a national problem, particularly in wetlands, which play a very important role in Senegal's agricultural production. Hence the need to invest even more in the subject (salinisation of water and soil), even if it should be stressed a return of rainfall (as in 2020) could slow the degradation process, with a drop in alkalinity and salinity.

Research prospects

Further research into species adapted to alkaline soils and saline irrigation water in the Niayes would be essential. This will enable abandoned super8cies to be recovered for better exploitation. Trials will be needed, with plant samples selected for proper field testing and regular, ongoing monitoring, in collaboration with accredited bodies such as ISRA. This will ensure a better choice of species for study. The Senegalese government, in collaboration with public and private organisations, has made several attempts to reclaim land, which in some areas (groundnut basin, eastern Senegal/upper Casamance) have produced more or less satisfactory results, through the construction of anti-salt dykes, reforestation campaigns or organic fertilisers, but the Niayes area needs to be better managed, as the situation is showing a negative trend in the light of the results obtained in this study.

The main problem facing the land is alkalinity. A situation that will not be easy to manage, given that the salinisation of the water table is mainly due to marine intrusion, encouraged by coastal erosion. Coastal erosion is a global problem, and statistics show that it is set to increase over time as a result of ongoing climate warming. The problem being the poor quality of irrigation water, research could be carried out into the solar drip method, which is an e.cient, economical and simple watering system that can help market gardeners in the Niayes area to increase yields while using ten times less water.

This research requires regular monitoring to check that the Kondenskompressor (the device) is working properly and that the plants are being watered properly. With proper experimentation, the Kondenskompressor method could be the solution to the use of saline water for irrigation. It's a method that needs to be tried out in certain areas of the Niayes region, such as Thiès and Louga, where there is a lot of sunshine most of the year.

References

Publications

Aissa. Nd (2001). *Farmers' organisations and strategies for developing pickling in the Fandéne Rural Community: examples from the villages of Darou-Khoudoss and Touba Ndiaye.* Master's thesis, UGB, 103 p.

Amadou. L. Nd (1995). *Étude et Cartographie des paysages de la Grande côte sénégalaise : Applications à la mise en valeur et à la conservation des ressources naturelles*. Thesis, FLSH, UCAD, 508 p.

Aminata. D. (June 2006). *Dynamique de l'occupation du sol dans les Niayes de la région de Dakar de 1954 à 2003 : exemples de la Grande Niaye de Pikine et de la Niaye de Yeumbeul.* UCAD DEA dissertation, 90 p.

ANDRE. P., Claude. E., Deliste, Jean-pierre, Reveret, SENE. A., BITONDON. D., ARISON. L. R. (2000). *L'évaluation des Impacts sur l'Environnement et Processus, acteurs et pratique dans le concours des prix du Ministre de l'Éducation du Québec.* Montin category Volume.

Aurélien C. (July 2007). *Sensitivity of specific coastal hydrosystems to millennial variations in sea level in Senegal.* Master 2 Sciences de l'Univers, Environnement, Ecologie 54 p.

Awa. B. (2007). *Les fonctions reconnues à l'agriculture intra et périurbaine (AIPU) dans le contexte dakarois ; caractérisation, analyse et diagnostic de durabilité de cette agriculture en vue de son intégration dans le projet urbain de Dakar (Sénégal).* Agro Paris Tech, UCAD 378p.

Ben Naceur. M., Rahmoune. C., Sdiri. H., Meddahi. ML, Selmi. M. (2001). *Effect of salt stress on germination, growth and grain production of some Maghrebian wheat varieties.* Sécheresse, 12 : 167-74.

Camara. MMB. (2008). *Impacts des aménagements sur les zones littorales : l'exemple de l'ouverture de la brèche sur la Langue de Barbarie (Grande côte du Sénégal).* Proceedings of a multidisciplinary international conference, 12p.

Centre for the Development of Horticulture Cambéréne-Dakar (1992). *List of varieties of market garden species for cultivation in Senegal.* ISRA, 40p.

Cheikh. B. (1960), *Le peuplement des Niayes du Nord (Sénégal), Dakar* (Diplôme d'études supérieures). Université de Dakar, 79 p .

Cissé I. (2000), *Utilisation des pesticides dans le système de production horticole dans la zone des Niayes : les produits et leur impact sur la nappe phréatique.* Thesis, UCAD, 187p.

Cissé. S. *L'analyse des stratégies de la gestion des ressources naturelles pour une agriculture durable (cas de la communauté rurale de Ndoga Babacar Département de Tambacounda).* Mémoire de 8n d'études.

Cissé. G., Odermatt. P., Maystre. L. Y., Tanner. M. (June 1999). Use *a GPS and GIS software to evaluate seasonal variations in the exploited areas marshalling sites in the urban fabric of Ouagadougou.* Sécheresse. V 10. N° 2.

Environment Code (2001). Law n° 2001-01 of 15-01-2001. 70 p.

CSE (Centre de Suivi Écologique). (2000). *Annuaire sur l'Environnement et les Ressources Naturelles du Sénégal.* Ministère de l'Environnement, Etat du Sénégal, Dakar, 268 p.

Dame. F. (Dec. 1990). *Modifications actuelles du climat et leurs conséquences sur l'environnement dans la région des Niayes Méridionales.* Master's thesis, UCAD, 100 p.

Edwige. B., Reij. C. (May 2009). *La transformation silencieuse de l'environnement et les systèmes de production au Sahel : Impacts des investissements publics et privés dans la gestion des ressources naturelles*. Permanent Inter-State Committee for Drought Control in the Sahel 61p.

Fama. T. (2002). *The dynamics of coastal lakes in Senegal*. Internship report, CSE, 36 p.

Faye Babacar (1997). *Les pratiques de gestion foncière dans la zone des Niayes : le cas de la Communauté rurale de Sangalkam*. Rapport de stage Université Gaston Berger 77 p.

Demoulin. D. (1970). *Etude géomorphologique du massif de Ndiass et de ses bordures*. Thèse de 3eme cycle, Département de géographie, FLSH UCAD 90 p.

Diallo. P. O. (1997). *Évolution des systèmes de production maraichères et durabilité des stratégies de gestion des terroirs dans la zone septentrionale des Niayes (secteur de Potou).* Master's thesis. Gaston Berger University. UFR de Lettres et Sciences Humaines. Geography Section.

Dieng M.N. (2008). *L'impact du maralchage dans la dégradation des ressources naturelles dans les Niayes de la bordure du lac Tanma*. Master's degree, Cheikh Anta Diop University, 100 p.

Dieng. M. (1984). *Protecting the environment for an even more beautiful Senegal*. Ed. Direction de l'Environnement, Ministère de la Protection de la Nature.

Midy document (1971). *Environnement et Pollution*, Tome I, n° 78, 37 p.

Document de synthèse LADA (2007). *Characterisation of agricultural production systems in Senegal*, 39p.

El Hadji. S. (2001). *Le quaternaire récent du Sénégal occidental (Lacs Retba et Tanma; estuaire de la Casamance). Implications eustatique et paléoclimatique des diatomées*. Thesis, 189 p.

Enda Ecopop, UN Habitats/CDS-MUAT. (2007). *Étude de l'économie de l'agglomération de Dakar*, 118 p.

GENDRAU. F., GUBRY. P. et VERON. J. (1996). *Populations et Environnement dans les pays du Sud*, Paris. Karthala, CEPED, 305 p.

Ibrahima. D. (2004). *L'ouverture d'une brèche à travers la Langue de Barbarie (Saint-Louis du Sénégal). The public authorities and the consequences of the breach*, 120 p.

Idrissa W. (2003). *Information et coordination dans les filières maraìchères au Sénégal.* DEA dissertation, 85 p.

Julien. H., Ndongo. F. D. (2006). *Caractérisation des sols cultivés dans le cadre de l'agriculture périurbaine de la région de Dakar, Sénégal - Evaluation of the salinity of cultivated soils associated with irrigation water in the context of peri-urban agriculture in Dakar, Senegal.* Master in Natural Environmental Sciences -University of Geneva.

Kaddour. D., Youcef. D., Gaouar. A., Beldjoudi. Z. (Oct-Nov-Dec 2003). *Secondary soil salinisation in the Sahara. Consequences for the sustainability of agriculture in new development areas. Science et changements planétaires.* Sécheresse. V 14, No. 4; 282 p. Cahiers d'études et de recherches francophones.

Lassalle. E. (1970). *Étude géomorphologique et bioogéographique de la région occidentale* *du lac Mbeubeussé (Presqu'île du Cap-Vert Sénégal),* Master's thesis, UCAD 125 p.

Lazar. A. (2009). *Impact de la variabilité climatique récente sur les écosystèmes des Niayes du Sénégal entre 1950 et 2004.* Thesis, 208 p.

Le Brusq. J.Y and Loyer. J. Y. (January 1982). *Évolution de la salinité des sols et des eaux en relation avec la riziculture submergée dans le Delta du =euve Sénégal.* O.ce de la recherche scienti8que et technique outre-mer. Centre de Dakar Hann à ORSTOM.

Mainquet. M. (1995). Professor at the University of Rennes. *Man and drought.* Masson géographie. 335 p.

Maimouna D. (2005) *Étude de la baisse du niveau de la nappe dans les Niayes du Gandiolais.* Master's thesis UGB, 124 p.

Mariama. D. (2005). *Conservation practices in the Niayes ecosystem of Pikine: the example of PASDUNE.* DEA dissertation, UNESCO Chair, UCAD, 72 p.

Mariètou. Nd (2012). *Environmental dynamics and management from 1970 to 2010 of wetlands in Senegal: a study of land use by remote sensing of the Niayes with Djiddah Thiaroye Kao (in Dakar), Mboro (in Thiès) and Saint-Louis.* Thesis, 371 p.

Mbaye. A. (1999). *Vegetable production in Dakar: importance, constraints and potential.* In Agriculture urbaine en Afrique de l'Ouest. Ed. O. Smith, International. Developpement research centre, Ohawa, 56-66. 59

Mor. G. (1989). *Etude et cartographie des milieux naturels à 1/50 000 : Coupure de Mboro (Louga ND-28-XX° Evaluation des ressources en eau dans la zone des Niayes* . UCAD Master's thesis, 110 p.

Mtimel. A., Hachicha. M. (Dec. 1995). *Salinisation and hydromorphy in Tunisian oases. Science et changements planétaires.* Sécheresse. V 6, N° 4 ; 374 p.

Niang. S. (2017). *Chemical and mechanical degradation of agricultural land in Gandiolais (northern coast of Senegal), analysis of current adaptation dynamics.* UGB thesis, 355 p.

Piraux. M., Bock. L., Bulgen. A., Lefort. E., Steyaert. P., Diagne. A. (1997). *In=uence of climatic, topographical and intensification conditions on the physical and chemical fertility of ferruginous soils in the Sahelo-Sudanian region.* Sécheresse. V 8. N°4.58 p.

Léona Rural Community Local Development Plan (2010).
With the support of the Millennium Village project, 76 p.

Local Development Plan (LDP) for the Malika district (2006).

Local Development Plan (LDP) Diender Rural Community. (2010). 129 p.

Local Development Plan (LDP) Ndiébène Gandiol rural community (2010). With technical support from GERAD, 80 p.

UNEP. (July 2006). *The Future of the Environment in Africa (past, present and future prospects).* Report, Steiner. M.

Pontanier, R. and A. M'Hiri. M'Hiri (1995). *L'Homme peut-il refaire ce 'il a défait?* ORSTOM, 460p.

Quenum. F. J. (1969). *Etude géomorphologique de la région de Cambéréne-Malika (Presqu'ile du Cap-Vert, Sénégal)*, TER, UCAD, 54 p.

Roger. B., NDiaye. K. M., Bassirou. K. (1994). *Alkalinisation/Sodisation, a danger for Sahelian irrigated perimeters*. Sécheresse. V 5. N°3.218 p.

Rokhaya. G. (2007). *Développement local et Problématique de la gestion du littoral dakarois à l'endroit de la Commune d'arrondissement de Fann-Point E-Amitié.* DEA dissertation, UNESCO Chair, UCAD, 77 p.

Rokhaya. Nd (1996). *Dynamique des groupements de promotion féminine dans la zone des Niayes (de Kayar à Fass Boye) et dans la gestion des ressources naturelles.* Master's thesis, UGB, 167 p.

Sall. M.M. (1971). *Dynamique et morphogenèse actuelle: contribution à l'étude géomorphologique du Sénégal Occidental.* Thesis, FLSH, UCAD, 291 p.

Serigne. F. (1995). *Hydrodynamic modelling of the nappes on the north coast between Cayar and Saint-Louis. Impacts des futurs prélèvements envisagés dans le cadre de l'approvisionnement en eau de Dakar et des environs.* Thesis, 102 p.

Thierno. Nd (2001). *Soil degradation in Senegal: the example of two rural communities (Kaymor and Mont Rolland).* Thesis, 313 p.

Wade C. T. (2003). *Ecosystem Environment. Gestion des usages littéraux au niveau de la grande côte sénégalaise : Analyse de l'impact de l'évolution de la ressource en eau sur les potentiels horticoles de la région des Niayes.* DEA dissertation. Cheikh Anta Diop University, Dakar.

Youcef. D., Amor. H. (Sept 1994). *Irrigation and salinisation in the Algerian Sahara.* Sécheresse. V 5. N°3; 218 p.

Youssou. Nd (1990). *Étude de l'évolution récente d'un écosystème intra urbain : cartographie des transformations des paysages des Niayes de Pikine-Thiaroye et environs.* Master's thesis, UCAD 87 p.

Ziza. FZ (1992) *Dynamics of water and salts under pivot irrigation in a Saharan environment. An attempt to optimise irrigation management.* Mémoire d'ingénieur INA, EL. Harrach, 101 p.

Articles, Conferences, Reports, Summaries and Manuals

Agence Nationale de la Statistique et de la Démographie (2010). "Situation Économique et Sociale du Sénégal en 2009". Ministry of Economy and Finance, 304 pp.

Ahmadou Moustapha Mbaye (Friday 1 April 2011). "Threats to the Niayes area and market garden basins / Isra launches filao regeneration project".

Ba. E. (2004). "Situation du Sénégal sur les questions d'érosion côtière". Planning workshop for the regional project for coastal protection and integrated coastal zone management, 15 p.

Barreto. S.P. (1963). "Études pédologiques des Niayes méridionales (entre Kayar et Mboro)". ORSTOM Dakar report. 120 p.

Bernard. F. et al (1996). "Contribution au rapport du groupe de travail sur les écosystèmes particulièrement vulnérables (Zones Humides continentales, Zones Humides littorales et milieu marin) dans le cadre du Plan National d'Action pour l'environnement (PNAE)". ORSTOM, 7p. UNESCO **Intergovernmental Oceanographic Commission**. (2001). "Des outils et des hommes pour une gestion intégrée des zones côtières; Guide Méthodologique" Volume II Manuels et Guide, 63 p.

UNESCO **Intergovernmental Oceanographic Commission**. (1997) "Guide Méthodologique d'Aide à la gestion intégrée de la zone côtière". Manuals and Guide 47 p.

Direction de la Prévision et de la Statistique (Senegal), "Situation économique du Sénégal1959-1979", 468 p.

Direction des travaux Géographiques et Cartographiques du Sénégal (DTGC). (2003), "Catalogue des produits", 15 p.

Dorsouma. Al H. and Mélanie R-D. (2009). "Climate variability, desertification and biodiversity: an integrated approach to adaptation", 18 p.

Bernard. F. et al (1996). "Contribution au rapport du groupe de travail sur les écosystèmes particulièrement vulnérables (Zones humides continentales, Zones Humides littorales et milieu marin) dans le cadre du Plan National d'Action pour l'Environnement (PNAE)", ORSTOM, 7p. **Dubois. J.** (1989). "Note sur l'ensablement de "Niayes" de Mboro et recherche d'un moyen efficace de lutte", ORSTOM, 8 p.

Lazar A., Garneau, Michelle; Lézine, Anne-Marie (2008). "Diachronic analysis of wetland evolution in the Niayes region of Senegal in relation to recent climate variability between 1950 and 2004". Université de Québec PP 121-149.

Le Meur. P. Y. (March 2004). "Note sur la problématique des sites de Thiès et Mboro. GRET, Ecocité partner. 8 p.

Manuel (October 2001). "Décentralisation et gestion des ressources naturelles; illustré sur les compétences des collectivités locales dans le domaine de l'environnement et de la gestion des ressources naturelles", DGL Felo with USAID/Senegal 28 p.

Mary. T.N. (2004). "Et Dieu sauva Saint-Louis. Le Canal de Délestage", 5 p.

Michel. P. (1965). "Notes complémentaires sur la région des Niayes de Cayar à l'embouchure du Sénégal", Bulletin-Mission-Aménagement-Sénégal-Saint-Louis No. 102. 6 p.

Pape. S. (2007) "Dakar et le littoral". UCAD, 4 p.

"Report of the United Nations Conference on Environment and Development (Rio de Janeiro, 3-14 June 1992).

Analysis **report**. (1993). "Étude sur l'Environnement urbain au Sénégal 4^{éme}Projet urbain du Sénégal Banque Mondiale", 250 p.

Sall, M. M. (2000). M. M. (2000) "Coastal erosion and coastal development". Reports n°2 Chaire UNESCO UCAD p 15-30.

Trochain. J.L. (1940). "Rupture de l'équilibre biologique dans les pays des Niayes (République du Sénégal)", p 57 - 60.

USAID and the Direction de l'Aménagement du Territoire. (1986). "Cartographie et télédétection des ressources naturelles de la République du Sénégal: Étude de la géologie des sols, de la végétation et des potentialités d'utilisation des sols". PP 144-147.

Wade. C. T. (Dec. 2010). "Onion transport and marketing network in the Niayes on Senegal's Grande Côte". UGB, RGLL, N°08 p 230-247.

Webography

http://www.lepopulaire.sn on 10 May 2010, November 2017.

http://www.sudquotidien.sn 10 May 2010, November 2017.

http://www.lequotidien.sn 12 May 2010, November 2017.

http://www.leral.net 15 May 2010, November 2017.

http://www.xibar.net 16 May 2010, November 2017.

http://www.journee-mondiale.com 19 June 2010, September 2017.

www.lagazette.sn 19 June 2010, September 2017.

http://www.seneweb.com 09 August 2010, September 2017.

http://www.seneweb.com 10 August 2010, September 2017.
http://www.lesoleil.sn. 05 November 2010, September 2017.
http://www.walf.sn. 07 November 2010, September 2017.
http://www.au_senegal.com. 03 December 2010, September 2017.
http://www.aps.sn 09 January 2011, October 2017.
http://www.environnement.gouv.sn 18 May 2011, October 2017.
http://www.ecocite.org 14 November 2011, October 2017.
Abdoulaye SIDY 25 May 2011. Salinisation des sols au Sénégal : sur les 3 800 000 ha cultivables plus de 1 230 000 ha sont affectés http://xalimasn.com consulted on 05 June 2011, October 2017.
Aissatou DOUCOURE 29 April 2011. Un tiers des superficies cultivables affecté par la salinisation au Sénégal www.xibar.net consulted on 08 June 2011, November 2017.
Aaron Akinocho 29 May 2014. 15,000 hectares of arable land saved from salinisation in Casamance www.agenceecofin.comconsulté 01 June 2014, November 2017.
Cherif Faye 31 March 2011. Menace sur la bande des Niayes. Djibo Ka sounds the alarm www.rufisquenews.com accessed 20 May 2011, November 2017.
Diakina Dia 26 March 2011. Menaces sur la zone des Niayes et cuvettes maraìchères / ISRA lance le projet de régénération du filao www.xibar.net consulted on 21 May 2011, November 2017.
Mathieu Ducrocq 23May 2011. Coping with risks in West Africa www.iucn.org consulted on 19 June 2011, December 2017.
Mitch 27 November 2009. Soil salinisation: a problem with multiple consequences http://casamance-passion.over-blog.com consulted on 12 December 2010, December 2017.
Tony Cissé 24 November 2012. The impact of soil salinity on vetiver growth http://vetiversenegal.blogspot.sn accessed 10 December 2012, December 2017.

Appendix 1: GROWTH OF DBS PLANTS IN DULL ENVIRONMENTS

Annexe 1 : CROISSANCE DES PLANTES EN MILIEU SALE

EFFETS TOXIQUES D'IONS DETERMINES

Comme l'a montré Harris (84), l'action nocive des sols n'est pas toujours proportionnelle à la pression osmotique du sous-sol salin. Selon l'espèce, chacun des divers éléments existant dans les solutions salées peut avoir un effet toxique spécifique sur la plante en plus de celui qu'on peut attribuer à la pression osmotique de la solution du sol. Les écrits traitant de l'effet toxique des ions sur la croissance des végétaux ont été revus par Magistad (144), Mibashan (154) et Hayward et Wadleigh (100). Les ions qui peuvent s'accumuler dans les sols salés sont : Na^+, Ca^{++}, Mg^{++}, K^+, Cl^-, SO_4^{-}, CO_3H^-, et NO_3^-.

Sodium

Bertrand (24) déclare que Na est toujours présent dans les tissus végétaux, mais parfois en quantités inférieures à 0,1 gr. par kg. de matière sèche. Cependant il apparaît que certaines plantes accumulent du sodium tandis que d'autres ont tendance à en éliminer.

Figure 4 - Non reproduite dans le document.

En se basant sur l'analyse de 77 espèces de plantes indigènes du New-Jersey et de 24 variétés de plantes cultivées, Wallace et ses collaborateurs (249) ont découvert que la teneur en Na des espèces indigènes variait de 0,00 à 2,04 % du poids sec, et les variétés cultivées de 0,00 à 3,00 %. Collander a trouvé (43) que la plupart des halophytes étaient très riches en Na, tandis que le sarrazin, le maïs, et le soleil étaient remarquables par leur manque presque total de Na. Hayward et ses collaborateurs (95) ont noté que les racines de pêcher peuvent accumuler dix fois la quantité de Na que l'on trouve dans les feuilles et les branches avec des valeurs de 0,36 % dans les petites racines; et Gauch et Wadleigh (74) ont noté de fortes accumulations de Na dans les racines et de faibles concentrations dans les tiges et les feuilles. Des effets toxiques spécifiques peuvent être provoqués par l'exclusion de Na quand se produisent en même temps des accumulations d'anions provenant du sous-sol, mais il ne semble pas que l'on doive classer ces cas comme relevant de la toxicité du sodium.

On a relativement peu d'observations permettant de préciser la toxicité spécifique de l'ion Na pour les plantes poussant dans les sols salins, mais Lilleland et ses collaborateurs (134) ont observé sur l'amandier, en Californie, une brûlure de l'extrémité des feuilles directement provoquée par la teneur en Na de la feuille. Cependant la solution du sol n'avait une teneur élevée ni en sel ni en Na, et le symptôme observé pourrait bien indiquer une nature alcaline du sol plutôt qu'une salinité. Ayers et ses collaborateurs (10) ont décrit deux types de brûlure de la feuille de l'avocatier qui ont été attribués à l'accumulation de Na ou de Cl. Les symptômes des dégâts dus à Na étaient des taches nécrosées ou écorchées près de la marge ou à la face interne de la feuille, tandis que dans le cas de dégâts dûs à Cl, la brûlure commençait à l'extrémité de la feuille et progressait vers la base du limbe et parfois le long des marges. Il est possible que l'accumulation de Na dans la plante soit associée à un abaissement de l'accumulation des autres cations au point de créer une balance cationique défavorable (Ratner 182, 183).

Calcium

L'ion Ca peut être toxique quand il s'accumule à fortes concentrations dans les solutions salines du sol, mais sa toxicité varie avec les espèces végétales. Par exemple Wadleigh et Gauch (243) ont montré que le guayule avait une tolérance relativement plus grande à un sous-sol salin produit par $CaCl_2$ qu'à ceux produits par d'autres sels neutres. D'autre part, Nasaewa (151) a publié que, dans cas des cultures de lin sur sol, des applications de $CaCl_2$ étaient plus toxiques que des applications de NaCl. Comme dans le cas de cultures additionnées de

$CaCl_2$, l'ion Cl s'accumulait aux niveaux supérieurs des plantes, et que Ca était lui aussi présent en fortes quantités, elle attribua la réaction de la plante à la toxicité du Cl et à un rapport Ca/K défavorable. Wadleigh et ses collaborateurs (245) ont effectué des recherches sur la composition minérale de Dactylis glomerata cultivé sur des sols additionnés de divers sels et ont trouvé que de très fortes concentrations de l'ion Ca dans la solution du sol peuvent être mortelles pour cette graminée quand l'anion associé est soit Cl soit NO_3. Comme l'addition de $(NO_3)_2Ca$ ne provoqua qu'une faible accumulation d'une petite quantité de Cl, ils conclurent que la toxicité du Cl n'était pas en cause. Lehr (132) dans ses recherches sur l'importance du sodium dans la nutrition de la plante a noté que la présence de Ca en quantités relativement fortes a un effet très nuisible sur la production du feuillage et des racines chez la betterave fourragère. La plante traitée au calcium est caractérisée par une couleur bleu-vert et une croissance rabougrie qui sont neutralisées par l'absorption de Na.

Gaugh et Wadleigh (72, 242) ont étudié l'absorption des éléments nutritifs par les haricots rouges nains cultivés sur sous-sols salins. Ils ont utilisé deux séries de cultures en solutions aérées, une série à laquelle on ajoutait NaCl, l'autre $CaCl_2$. En plus de la culture nutritive de base (de pression osmotique 0,5 atm.) on ajoutait dans chaque série à la solution nutritive de base quatre doses de sels de pressions osmotiques de 1, 2, 3 et 4 atm. Ils ont trouvé que les plantes des séries à $CaCl_2$ avaient une croissance moins rapide que celles des séries à NaCl, et ont attribué la différence surtout à une croissance des racines plus médiocre dans les séries à $CaCl_2$. Les plantes où NaCl, où Na prédominait dans les racines, absorbaient au total de plus fortes quantités de N, P, et K que les plantes à $CaCl_2$. Ils suggèrent que l'action différentielle des ions Na^+ et Ca^{++} sur l'hydratation des colloïdes peut entrer en ligne de compte et que l'effet inverse de Na par rapport à Ca sur la perméabilité des cellules aux sels peut être un facteur de première importance pour expliquer les résultats obtenus.

Magnésium

On a reconnu que de fortes accumulations de Mg dans le sous-sol étaient toxiques pour les plantes en plus de toute inhibition liée à la pression osmotique : Sigmond (207), Trealease et Trealease (288), et Wadleigh et Gauch (243). Ces derniers ont trouvé que le guayule était tué quand on ajoutait à une solution nutritive assez de $MgCl_2$ pour produire une pression osmotique de 1 atm., tandis

que dans des conditions semblables les haricots rouges accusaient une réduction de croissance d'environ 10 % seulement. Trealease et Trealease (228) déclarent que les dégats causés au blé par le Mg varient surtout en fonction du rapport Mg/Ca, bien que ces dégâts puissent être influencés par d'autres produits chimiques; et Gauch(70) associait la nocivité du Mg à un apport insuffisant de Ca. Ratner (182) indique que des quantités de Mg remplaçable supérieures à la normale créent un équilibre Ca-Mg défavorable, et cite les résultats obtenus par des chercheurs russes qui précisent que lorsque Mg constitue 50 à 60 % de la capacité d'échange en cations, il n'y a pas de diminution de rendement; mais au dessus de ce taux il peut y avoir une diminution nette de la croissance.

Potassium

Des accumulations de K dans la solution du sol sont peu fréquentes, mais peuvent se produire. Dans ce cas, il est peu vraisemblable que l'on observe des effets inhibiteurs spécifiques de la réaction de la plante dus à un excès de K si cette accumulation est partiellement compensée par Ca. On a noté des cas où des quantités relativement fortes de K ont produit des symptômes de chlorose du fer - Walsh et Clarke (250) -. Boynton et Burrel (30) ont décrit des taches sur les feuilles du pommier dues à une déficience en Mg qu'ils ont attribuées à l'effet antagoniste d'une forte teneur en K à la surface des racines ou dans l'arbre. Miyake (158) déclare que : "les sels de potassium, de magnésium ou de calcium constituent des poisons pour le riz quand on les applique séparément, mais l'effet nocif disparaît plus ou moins complètement quand on les mélange en proportions convenables."

Il est évident que les cations diffèrent de façon marquée dans leur effet sur les propriétés physiques des constituants colloidaux du protoplasme. Des effets antagonistes se révèlent entre diverses paires de cations, par exemple entre Ca^{++} et Mg^{++} ; et entre des cations monovalents et divalents - Lundegardh (137). Il n'entre pas dans le but du présent rapport de passer en revue la question de l'antagonisme, mais de nombreux chercheurs ont noté que la croissance de la plante est réduite par une proportion défavorable des ions - Collander (43) et Osterhout (167, 168).

Chlorures

L'effet toxique de l'ion Cl a été noté par de nombreux chercheurs, mais ils ont eu parfois tendance à considérer que la toxicité du Cl était synonyme de l'effet défavorable d'une forte salinité du sol. Toutefois, pour de nombreuses espèces, on a observé que les sels de Cl n'inhibent pas davantage la croissance que des concentrations isosmotiques de sels de SO_4 - Eaton (55), Hayward et Long (93), Magistad et ses collaborateurs (145). Eaton a découvert que les boutures de citronier, les haricots "navy" (Phaseolus vulgaris) et le sorgho nain, cultivés sur sable, étaient plus sensibles à Cl qu'à SO_4 à pressions osmotiques égales, tandis que les tomates et les betteraves à sucre avaient une meilleure tolérance au Cl. Harper (83) signale que le pacanier est gravement endommagé quand la teneur en Cl des feuilles dépasse 0,6 % et a observé des dégâts sur les feuilles de pacanier, d'orme et d'hickory quand la dose de Cl du sol sec dans la zone des racines est supérieure à 200 p.p.m. Cooper et Gorton (44) ont découvert que la brûlure des feuilles est associée à l'accumulation de Cl chez l'avocatier, l'arbre à pamplemousse, la sapote blanche et l'Eugenia. Le taux de Cl des feuilles d'avocatier présentant des brûlures de l'extrémité allait de 0,22 à 1,48 % du poids sec. La brûlure des feuilles de pamplemousse se produisait plus rarement à des concentrations inférieures à 0,7 % de Cl et les feuilles les plus sévèrement brûlées avaient habituellement une concentration d'environ 2,0 %.

Hayward et ses collaborateurs (95) ont découvert, à la suite d'une étude de 5 ans sur le pêcher que, à pressions osmotiques égales, les solutions chlorurées entraînaient une plus forte réduction de la croissance des végétaux et étaient plus toxiques que les solutions sulfatées. Une réduction de croissance et de production due à une augmentation progressive de la concentration de NaCl dans la solution du sol a été signalée par Wadleigh et Ayers (238) et Bernstein et Ayers (22) pour les haricots; par Ravikovitch et Bidner (184, 185) pour la vigne et le trèfle; par Heller et ses collaborateurs (105) pour la tomate; par Millington et ses collaborateurs (155) pour l'avoine et l'orge; par Doughty et Stalwick (50) pour la luzerne, l'avoine et le blé.

Garner et ses collaborateurs (69) en travaillant sur le tabac ont fait des recherches sur la toxicité du Cl. Lorsqu'il y avait dans les feuilles de tabac une forte accumulation de Cl à la suite de l'emploi d'une dose massive de KCl comme engrais, ils ont trouvé une réduction prononcée de la teneur en acide malique dans les feuilles, et ils ont noté également que l'activité amylolytique normale était troublée et que les feuilles se gorgeaient d'amidon. Baslavskaja (18) a observé que l'accumulation de l'ion Cl dans les feuilles de pomme de terre gênait la photosynthèse en provoquant une réduction de leur teneur en chlorophylle, même s'il y avait une augmentation du rapport amidon-sucre. Schuphan (198) a dit qu'il n'était pas possible de généraliser l'effet du Cl sur le métabolisme des hydrates de carbone car les espèces ont à l'ion Cl des réactions très variées.

L'augmentation de pression osmotique du sous-sol peut avoir pour effet de réduire la croissance et la production d'une plante cultivée sans provoquer des symptômes spécifiques de lésion (13, 55). Dans ce cas, il peut y avoir une réduction de la taille des feuilles, de la longueur et du diamètre des tiges résultant d'une inhibition de l'activité des méristèmes primaires et secondaires - Hayward et Long (93, 94). Dans quelques exemples, de fortes accumulations de Cl ont tendance à augmenter l'épaisseur et la succulence des feuilles (93, 232, 255). Boyce (29) a étudié l'hypertrophie due au sel des plantes des dunes et a reconnu qu'elle était liée à la tolérance de la plante à d'importants dépôts de sel résultant de la pulvérisation de sel sur les feuilles. Quand elles étaient soumises à de fortes concentrations de sel, les espèces les plus tolérantes étaient caractérisées par une plus grande succulence et une capacité à accroître l'épaisseur du mésophylle.

Dans de nombreux cas, on peut observer des signes évidents de dégâts dus au Cl. Le plus commun est une brûlure de l'extrémité, roussissement ou nécrose du tissu de la pointe de la feuille qui peut s'étendre le long de la marge et plus tard intéresser de la moitié aux 2 tiers et dans certains cas la surface entière de la feuille. Ceci peut être précédé par un début de chlorose et de jaunissement de la feuille. Dans le cas de brûlures graves, la chute de la feuille se produit. La mort des petites branches est aussi un symptôme habituel, la croissance des racines peut être ralentie et il peut y avoir une réduction de la production et de la taille des fruits (22, 44, 126, 184).

Sulfates

Les observations faites sur un grand nombre de plantes cultivées indiquent que de fortes concentrations de l'ion SO_4 sont toxiques. Ceci a été signalé pour

WS/112.52

le lin (99), la tomate (55), Dactylis glomerata (245), le poireau (198), le citronier (81). Hayward et Spurr (99) ont trouvé que, à concentrations osmotiques égales, SO_4Na2 inhibait davantage la production de graines de lin que NaCl ou $CaCl_2$; et Wadleigh et ses collaborateurs (246) rendent compte de résultats semblables pour le guayule (Parthenium argentatum). Eaton (55) a trouvé que, en milliéquivalents SO_4 paraissait être moitié moins toxique de Cl pour certaines plantes. Haas et Thomas (81) ont observé la marbrure des feuilles de citronier avec une coloration jaune ou bronzée, quelques brûlures et une chute accentuée des feuilles quand il y avait une concentration importante de SO_4 dans la solution du sol.

Harris et ses collaborateurs (89,90) ont noté des différences concernant l'accumulation des ions Cl et SO_4 dans les liquides cellulaires des feuilles de différentes variétés de coton. Les variétés égyptiennes de coton ont tendance à accumuler beaucoup plus de Cl dans leurs tissus que les variétés américaines à fibres courtes, et la même tendance a été observée en ce qui concerne SO_4. Lyon (138), travaillant sur deux races de tomates et sur la génération F_1, a constaté que la race Red Currant était moins touchée par de fortes concentrations de SO_4Na_2 que les deux autres en ce qui concerne le poids sec des racines, le poids frais des fruits et le poids moyen des fruits mûrs.

Bien qu'on n'ait que des renseignements incomplets sur les raisons pour lesquelles l'ion SO_4 a des effets inhibiteurs sur la croissance de certaines espèces, il est évident que de fortes concentrations de SO_4 peuvent limiter l'activité de l'ion Ca dont dépend l'absorption cationique de la plante. Des analyses de feuilles de haricot (74), de pêcher (95) et de Dactylis glomerata (245) ont montré que les tissus contenaient moins de Ca et plus de Na et de K lorsque SO_4 était l'anion prédominant dans le sous-sol que lorsque c'était Cl. Ces résultats font penser que les effets spécifiques défavorables de SO_4 peuvent être dus à la perturbation d'un équilibre cationique favorable dans la plante.

Bicarbonates

On pense que l'ion CO_3H est très toxique pour les plantes, mais il est très difficile de mettre ceci en évidence à cause des relations entre l'ion CO_3 et le pH; Breazeale (32) déclare que CO_3HNa se trouve rarement en concentrations toxiques dans les sols alcalins et que sa toxicité exprimée par rapport à Na est à peu près la moitié de celle de CO_3Na2. Heller et ses collaborateurs (104) ont étudié les tomates en culture sur sable en utilisant NaCl et CO_3HNa en concentrations de 100 à 3000 p.p.m. Les résultats qu'ils ont obtenus indiquent que CO_3HNa était plus toxique que NaCl, et qu'il provoquait une inhibition très nette de l'absorption de Ca par les plantes. Harley et Lindner (82) ont indiqué que des vergers de pommiers irrigués avec de l'eau relativement riche en CO_3H avaient tendance à devenir chlorotiques; et ils ont trouvé que la situation pouvait être améliorée en irriguant ensuite avec de l'eau pauvre en CO_3H. Ils ont observé aussi une forte incrustation de CO_3Ca et CO_3Mg sur les racines des arbres des vergers qui avaient été irrigués pendant un certain temps avec de l'eau riche en bicarbonates de Ca et Mg. Ils pensent que ces dépôts peuvent produire des symptômes de chlorose en agissant sur la nutrition minérale de l'arbre.

Gauch et Wadleigh (72) ont trouvé la preuve d'une nette spécificité de la tolérance à l'ion CO_3H dans leurs études sur deux graminées, Chloris gayana et

Paspalum dilatatum. L'addition à une solution nutritive de 12 milliéquivalents par litre de CO_3H n'avait aucun effet sur la croissance de Chloris gayana, mais était extrêmement toxique pour Paspalum dilatatum, provoquant une grave chlorose ou la mort de la plante. Wadleigh et Brown (239, 240) ont étudié l'effet de CO_3H sur les haricots rouges nains et sur les betteraves rouges potagères, en utilisant des solutions nutritives auxquelles on avait ajouté 8, 16 et 32 milliéqu./l. de CO_3HNa. Toutes les solutions étaient préparées au pH 8,0. Les résultats obtenus montrent la différence des réactions à CO_3H suivant les différentes espèces de plantes. Chez les haricots, les fortes concentrations de CO_3H entraînaient une inhibition marquée de la croissance et une chlorose prononcée. Les traitements entraînèrent un ralentissement de l'"activité" du fer, un abaissement du taux de Ca des feuilles et une augmentation de leur taux en K. La croissance des betteraves était moins influencée par l'ion CO_3H que de nombreuses espèces étudiées. Des doses croissantes de CO_3H ne provoquaient qu'une diminution relativement faible de la croissance avec peu de signes macroscopiques de chlorose, bien qu'il y ait eu une certaine diminution de la concentration des feuilles en chlorophylle, accompagnée d'une augmentation de la teneur en bétanine. Il y avait une accumulation très nette de Na et d'oxalate.

Steward et Preston (212) ont effectué des recherches concernant l'action de l'ion CO_3H sur l'absorption et le métabolisme du bromure en utilisant des rondelles de pomme de terre dans des solutions aérées. A pH constant, l'augmentation de la concentration externe de CO_3HK diminuait à la fois la synthèse des protéines et l'accumulation de bromure. Des preuves indirectes montraient que CO_3HK diminuait aussi la respiration et le métabolisme des hydrates de carbone. Les études citées montrent que l'effet défavorable de CO_3H sur la réaction de la plante est dû à l'absorption et au métabolisme des ions et qu'il peut y avoir des différences marquées dans la toxicité de CO_3H suivant les espèces végétales.

Nitrates

L'ion NO_3 peut s'accumuler à des taux assez élevés dans certains sols naturellement salins et cet état de choses est caractérisé par le développement de "dépôts de nitre" (101, 195, 215). Delley (125) a passé en revue les théories concernant la formation de ces dépôts et déclare que "aucune explication vraiment satisfaisante n'a été donnée jusqu'à présent sur l'origine des énormes dépôts de nitrates au Chili ou dans d'autres endroits." Larson (130) a décrit un sol à salinité alcaline dans la vallée de Yakima, état de Washington, dans lequel NO_3 constituait 55 % des anions dans le profil du sol jusqu'à une profondeur d'un mètre environ, et 21 à 63 % des anions dans l'eau de la nappe phréatique.

Il est habituellement difficile de faire une distinction nette entre un effet spécifique de l'ion NO_3 et les effets concomitants produits par la pression osmotique du sous-sol ou les cations complémentaires, mais on connaît plusieurs exemples d'inhibition de croissance produite par de fortes doses de NO_3. Chapman et Liebig (41) ont constaté que la croissance des plantules d'oranger était considérablement ralentie dans des solutions contenant 840 p.p.m. d'azote nitrique. Leonard et ses collaborateurs (133), utilisant des solutions nutritives d'azote, ont dit que les feuilles de la patate douce devenaient vert sombre et que quelques-unes d'entre elles tombaient lorsqu'on appliquait la solution à 200 p.p.m. Les feuilles produites avec la solution à 800 p.p.m., et surtout avec celle à 1600 p.p.m., étaient chlorotiques, et ces dernières mouraient avant la fin de l'expérience. Le poids frais de l'appareil végétatif, racines comprises, diminuait de façon nette à la dose de 400 p.p.m. d'azote. Eaton et Rigler (57) ont cultivé du coton à des taux d'N allant de 14 à 396 p.p.m. et ont constaté que de fortes concentrations de NO_3 réduisaient la croissance végétative et le nombre de capsules formées. Les plantes cultivées au taux d'N le plus élevé étaient moins grandes d'environ 25 % que celles cultivées aux taux intermédiaires (56 et 224 p.p.m.). L'observation ancienne de Headden (101), qui rapportait que l'accumulation de NO_3 dans le sol provoquait la production de betteraves à sucre de qualité inférieure à cause d'une faible teneur en sucre, a été vérifiée de nombreuses fois.

La capacité qu'ont une espèce végétale ou une variété données de germer et d'établir leur plantule est souvent le facteur limitatif en matière de production agricole; et, en milieu salé, cette phase du cycle vital de la plante peut être la phase critique. Les sols et les eaux salés peuvent agir de deux façons sur la germination :

(1) il peut y avoir assez de sel dans le milieu où se trouve la graine pour augmenter la pression osmotique de la solution du sol au point de retarder ou d'empêcher l'absorption de l'eau nécessaire, et

(2) certains sels ou ions qui s'y trouvent peuvent être toxiques pour l'embryon et la plantule.

Harris (84), Magistad (144) et Uhvits (229) ont passé en revue les écrits publiés sur cette question et soulignent que dès 1896, des chercheurs du Wyoming et d'ailleurs travaillaient sur la question de l'action d'une forte pression osmotique de la solution du sol (34, 35, 208, 214). Leurs études montrèrent que la germination était retardée par les sels en solution et que l'importance du retard de la germination et des dommages aux graines et plantules était en raison directe de la pression osmotique de la solution saline. Avec la plupart des substances en solution employées , l'augmentation de pression osmotique provoquait une diminution du gonflement des graines. Si la pression osmotique était suffisamment élevée, aucune germination ne se produisait; mais on remarqua que les diverses espèces de plantes agricoles montraient des tolérances aux sels différentes pour une concentration saline déterminée. Stewart (214) remarqua que les céréales avaient une tolérance aux sels supérieure à celle des légumineuses et établit la liste de leur tolérance aux sels relative dans l'ordre décroissant suivant : orge, seigle, blé, avoine. Pour les légumineuses, il donnait l'ordre de tolérance suivant : pois, trèfle rouge, luzerne, trèfle blanc. Harris (84) fit des essais de germination sur un grand nombre de plantes cultivées pour des concentrations de sels allant de zéro à 10.000 p.p.m. (1 % du poids sec), en utilisant plusieurs sels, isolément ou en combinaison. Il constata de grandes différences entre les plantes en ce qui concerne la germination en sous-sol salin et établit une liste des plantes étudiées dans l'ordre de tolérance décroissante suivant : orge, avoine, blé, luzerne, betterave à sucre, maïs, pois fourragers.

Des études plus récentes ont confirmé ces premiers travaux (9, 12, 159, 193, 194, 205, 229). Shive (205) employa une méthode de culture sur sable, sel par sel, pour étudier la germination des haricots et du maïs à des pressions osmotiques de 0,5 à 8,0 atm., et conclut que "le retard de germination est directement lié à la quantité d'eau absorbée par les graines, qui, à son tour, dépend de la concentration des solutions du sol". Rudolfs (193, 194) a utilisé des graines préalablement trempées qu'il faisait germer sur papier filtre avec un seul sel à des pressions osmotiques allant jusqu'à 7 atm. Sauf dans le cas de quelques-unes des solutions plus faibles, la germination et la croissance des racines diminuaient en fonction inverse de la concentration des sels. Millington et ses collaborateurs (155) ont poursuivi en Australie occidentale des études de germination contrôlée en sols salins sur l'avoine, l'orge, le blé et le seigle, pour des concentrations en sels allant de 0,10 à 0,30 % de NaCl. Ils ont constaté que le seigle avait la meilleure germination pour les conditions de salure les plus accentuées, mais qu'il avait une tolérance relativement médiocre durant la suite de sa croissance et de son développement. Les orges fourragères montrèrent la plus forte capacité à la fois pour la germination et pour la croissance d'hiver en sols salins.

L'évaluation du taux de salinité conditionnant la germination des graines sur le terrain est difficile, car l'humidité du sol et la concentration en sels changent continuellement à cause de l'évaporation, des mouvements capillaires de l'eau, des pluies ou de l'irrigation. Ayers et Hayward (12) ont décrit une méthode de laboratoire pour mesurer les effets de la salinité du sol sur la germination qui comporte l'humidification et la salinisation d'un sol non salin, de façon à obtenir un pourcentage d'humidité du sol et un taux de salinité déterminée. On détermine le taux d'humidité du sol et la richesse en sel de l'extrait de saturation, et ces données permettent de calculer la pression osmotique de la solution du sol dans l'essai de germination. Des quantités de sol préconditionné sont placées dans de grands plats de culture après avoir été pesés; on plante un nombre déterminé de graines et les cultures sont placées dans une pièce à température constante pour éliminer la variable température. Cette façon de procéder réduit aussi à un minimum la variation de l'humidité du sol, ce qui est important en matière de germination de graines (48, 113).

Avec cette technique, Ayers et Hayward (9, 12) ont étudié la luzerne, l'orge, les haricots rouges, le maïs, les oignons et les betteraves à sucre à des taux de salinité allant de 0,05 à 0,4 % du sol sec. Les pressions osmotiques des solutions du sol, calculées d'après la conductivité électrique de l'extrait de saturation et le taux d'humidité du sol au moment de l'ensemencement, allaient de 0,7 à 25,3 atm. Aucune graine n'a germé au taux de 0,4 %, mais pour l'orge (California,Marrioutt) on obtint 80 % de germination au taux de 0,3 % de sel (pression osm. de 20 atm.). On eut avec la luzerne 80 % de germinations avec 0,1 % de sels ajoutés (pression osm. 7,3 atm.), mais la germination des betteraves à sucre était réduite de 50 % au taux de sel de 0,08 % (pression osm. de 5,8 atm.)

Appendix 2:

MAP COORDINATES OF THE AREA		
PTS	X	Y
1	22765 6	1631207
2	23801 3	1620628
3	26897 6	1613767
4	29706 2	1611949
5	38940 2	1752282
6	34392 4	1778758

NIAYES ZONE
AREA = 7804512700 m2
LENGTH = 187 km as the crow flies
WIDTH = 60 KM as the crow flies

GEOGRAPHICAL COORDINATES OF THE NIAYES AREA

PTS	LATITUDE	LONGITUD E
1	14°44'27 "N	17°31'46 "W
2	14°38'47 "N	17°25'56 "W
3	14°35'14 "N	17°08'40 "W
4	14°34'23 "N	16°53'01 "W
5	15°50'50 "N	16°01'59 "W
6	16°05'02 "N	16°27'33 "W

Source: DTGC, 2011.

Appendix 3: Photographs of Lake Wouye in 2007 before its transformation into a retention basin

Photos of ceans at Lake Wouye in 2007

Source: M. Fall July 2007

Watering the plantations with a motor pump. There were only a few in the area in 2007.

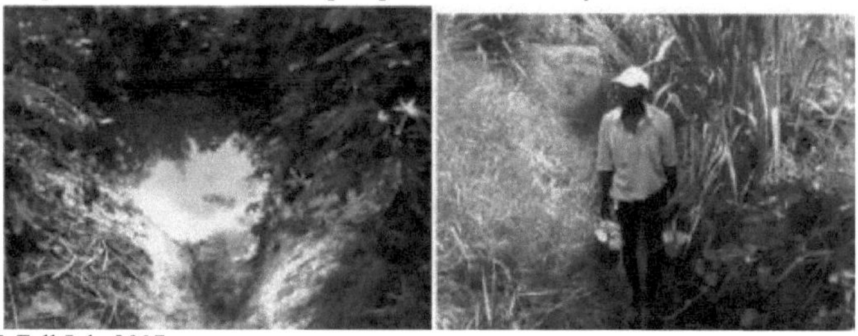

Source: M. Fall July 2007

Watering the plantations with watering cans was much more difficult than with the motor-driven pump, requiring several trips to and from the hanes.

Photos taken at the Saed laboratory in 2010

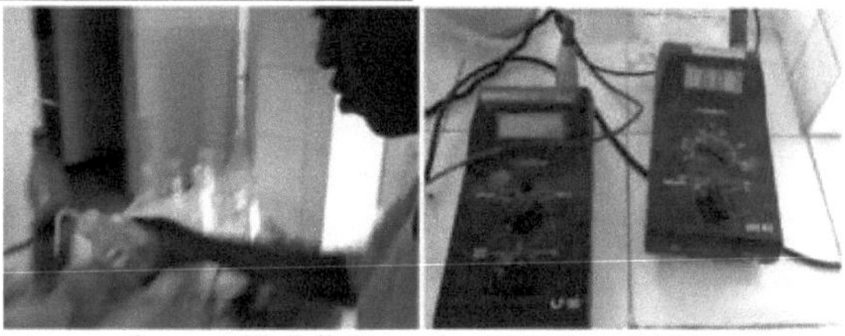

Source: M. Fall, December 2010 Ross Béthio

Analysis of soil and water samples at the SAED soil laboratory in Ross Béthio.

Appendix 4: Interview guide

MAINTENANCE GUIDE FOR MARKET GARDENERS

1. How many market garden plots are there, and what is the total area?
2. How much arable land has been lost to salinisation (village, lake), and when did you notice this phenomenon?
3. What were the yields before salinisation?
4. What are the current yields?
5. Is salinisation increasing over the years, and if so, are there any ways of combating it?

Appendix 5: Table of purchase and selling prices for market garden produce

Market garden products	Purchase price	Selling price
Leek	1500 to 5000 F /board	100 to 500 F/kg
Cabbage	1500 to 5000 F/board	100 to 500 F/kg
Tomato	3,000 to 10,000 F/case	100 to 300 F
Green onion	2500 to 15 000 F/bag of 30 and 35kg	100 to 400 F/kg
aubergine	7500 to 25 000 F/bag from 50 to 55kg	100 to 300 F/kg
Chilli	100 to 750 F/kg	125 to 800 F/kg
Radish	50 to 300 F	100 to 400 F
Bissap	10,000 to 20,000f/lot	100 to 150 F
Green chilli	150 to 300 F	200 to 400 F
Nana	5000F/board	25 F
Carrot	200 to 300 F	250 to 400 F
Navée	100 to 200 F	150 to 250 F
Potato leaf	25,000 F to 50,000 F/field	25 F
cucumber	100 to 200 F	200 to 300 F
Income per day	500 to 20,000 F	-
Income per year	360,000 to 7,200,000 F	-

Source:M. Fall, 2017

Annex 6 : Species grown in the Niayes area

Maraîchage	Aubergine (Solanum melongena L.) Gombo (Abelmoschus esculentus) Jaxatu (Solanum aethiopicum L. Kamba) Strawberry (Fragaria vesca L.)
Basil (Ocimum basilicum L.) Common Nana (Mentha sp.) Nana menthe (Mentha sp) Green cabbage (Brassica oleracea L. var. Capitata) Chilli pepper (Capsicum frutescens L.) Pepper (Capsicum annuum L.) Bissap (Hibiscus sabdariffa L.) Lettuce (Lactuca sativa L.) Courgette (Cucumis pepo L) Carrot (Daucus carota L.) Green bean (Phaseolus vulgaris L) Cassava (Manihot esculentum) Table tomato (Lycopersicum esculentum var.) Cherry tomato (L. esculentum var. cerisae) Turnip (Brassica napus L.) Melon (Cucumis melo L.) Watermelon (Citrillus colocynthis L. Schrad) **Source: M. Fall, 2017**	Potato (Solanum tuberosum L.) Cucumber (Cucumis sativus L.) Asparagus (Asparagus officinalis) Pumpkin (Cucumis maxima L.) Onion (Allium cepa L.) Leek (Allium porum L.) Squash Sweet potato (Ipomea batatas L.) Garlic (Allium sativum L.) Fennel Parsley (Petroselinum criapum L.) Cowpea (Vigna unguiculata L. Walp.) Chives (Allium schoenoprasum)

Appendix 7 Irrigation at Diender

L'irrigation des plantes se fait manuellement, soit à l'aide de seau en plastique, soit grâce à un tuyau souple relié à une moto-pompe pour l'exhaure de l'eau. L'eau est déversée directement au pied des plantes.

Comme fumure, une poignée d'urée a été appliquée à chaque plant, juste après le repiquage. Il est prévu plus tard, l'épandage de fumure organique ou de tout autre engrais disponible pour le paysan.

Les semences de tomate sont auto-produites, donc prélevées sur la récolte précédente. Elles donnent des fruits de même calibre que ceux de la variété Xeewel i Nawet, sélectionnée au C.D.H., c'est-à-dire intermédiaire entre la grosse tomate et la petite, du genre cerise, d'où le nom "Duma Dara" donné à cette tomate.

La culture de tomate vient dans la rotation des cultures après celle de la pomme de terre. Cette succession de deux espèces appartenant à la même famille des Solanacées sur la même parcelle, présente des risques sur le plan phytosanitaire (par exemple : présence de nématodes), de même que celui de l'épuisement du sol par la première culture.

Le seul prédateur signalé sur la tomate est Heliothis armigera qui perce les fruits et provoque la chute des fleurs. L'association Décis-Lannate était utilisée pour la lutte contre cet insecte nuisible. Il est à remarquer que ces deux produits pris séparément, assurent un excellent contrôle des populations du ravageur. En outre, il a été observé quelques plantes virosées.

Les principaux problèmes qui ressortent de cet expose sont :

- l'état phytosanitaire des semences : le prélèvement des semences sur la production augmente les risques de prolifération de certaines maladies virales, dont très souvent le maraîcher ignore l'existence, aussi un effort devra être fait dans le sens d'une meilleure information sur le choix des semences ;

- l'utilisation des fertilisants est un acquis, mais leur disponibilité reste problématique. En plus, les doses requises et les moments d'épandage ne sont pas bien maîtrisées ;

- la position de la culture de tomate dans la succession culturale après la pomme de terre, n'est pas la plus indiquée.

Irrigation

La baisse de la nappe phréatique, conséquence de la faiblesse de la pluviométrie des années précédentes, a conduit à la désertion des bas-fonds pour les versants dunaires.

Ce déploiement des cultures vers la périphérie des Niayes pose certaines difficultés. L'arrosage est devenu une nécessité et la fréquence de celle-ci est plus rapprochée, d'où un accroissement de la main-d'oeuvre. L'utilisation de moto-pompes pour l'exhaure de l'eau, permet une réduction de la main-d'oeuvre, mais augmente les charges par des frais de carburant, de lubrifiant et de maintenance, qui très souvent, fait défaut pour cette dernière. Les céanes de plus en plus profondes, n'offrent qu'une faible quantité d'eau dans la journée, ce qui entraîne une réduction des superficies cultivées. La pauvreté des sols en éléments fertilisants, impose l'emploi de fumure organique et minérale pour avoir un niveau correct de production, d'où une augmentation des charges de fonction de l'exploitation.

Annex 8 : Cultivation techniques for a number of market garden products

Pomme de terre

Variété : Claustar

Techniques culturales

Nombre de tiges m^2 brut : 8 à 10

Quantité de semences utilisée ± 1.000 kg/ha

Profondeur de plantation : 15 cm

Pas de buttage

Densité de plantation : ± 7 morceaux de tubercules par m^2 brut ;
5 à 6 morceaux par tubercule.

Manioc

Culture de manioc chez El Hadji Matar Diop (± 9 ha

Expérience dans le secteur	:	20 ans
Problèmes rencontrés	:	cochenilles, mosaïque, sécheresse
Variété	:	Kombo (type clair et type foncé)
Caractéristiques variétales	:	le type clair est plus résistant à la cochenille et plus tolérant à la sécheresse. Malheureusement, il est extrêmement sensible à la mosaïque
Plantation	:	juin (au moment des premières pluies)
Récolte	:	fin juillet/août
Prix de vente au bana-bana		15.000 F/sac (50 kg) (les prix les plus élevés seraient obtenus en mai/juin).

Cependant, la culture sur les versants dunaires longtemps souhaitée par le C.D.H., présente un intérêt surtout pour la pomme de terre. Les tubercules récoltés ne seront plus gorgés d'eau, comme ceux récoltés dans les bas-fonds, auront une meilleure qualité gustative et se conserveront plus longtemps, avant leur écoulement vers les marchés.

Association gombo-oignon

Culture d'oignon en association avec le gombo

Durée des pépinières	:	1 mois
Semis du gombo	:	1/2 mois après repiquage de l'oignon
Récoltedes bulbes	:	2 mois après repiquage
Variété	:	Texas Early Grano (le Jaune hâtif de Valence est utilisé en culture tardive).
Densité moyenne	:	38.400 pl/ha
Achat des semences dans le secteur maraîcher	}	1.250 F/100 g pour le gombo 6.000 F/100 g pour l'oignon
Vente des bulbes	:	300 F/kg pour les récoltes hâtives 75-100 F/kg pour les récoltes en avril/mai
Engrais utilisés	:	14-7-7, coques d'arachide, fumier.

Appendix 9: Production levels of the main crops by region (2004/05 in tonnes)

Product	Regions in the intervention area				Other regions	National
	Dakar	Thiès	Louga	SaintLouis		
Onion	8 750	2 750	10 625	12 750	5 125	40 000
Tornate	4 550	3 650	3 400	11 000	2 375	25 000
Cabbage	6 200	8 900		275	1 625	17 000
Potatoes	4 894	2 000	428			7 282
Okra (abermoscus)	2 832	1 416			3 652	8 500
Bitter aubergine	5 487	3 990			3 082	12 479
Aubergine	1 134	851		567	851	3 403
Green bean	2 500	1 250	500	250		4 500
Chilli	1 955	788		575	1 145	4 463
Carette	2 125	3 251	133	350	2 575	8 506
Turnip	7 048	2 493		431		9 972
Sweet potato	ND			26 485		26 485

Source: Agricultural Policy Unit, Ministry of Agriculture

Appendix 10: Indicator of the normal amount of fertilisation for some market vegetable crops

Culture	Bottom dressing	Maintenance manure
Onion	10-20 t/ha of organic matter 250 kg/ha of 10-10-20	-
Carrot	10-20 t/ha of organic matter 500 kg/ha of 10-10-20	300 kg/ha from 10-10-20 to 20 and 40 days after sowing
Cabbage	20-30 t/ha of organic matter 250 to 300 kg/ha of 10-10-20	200 kg/ha of 10-10-20 after 20 and 35 days of planting
Tomato	20-30 t/ha of organic matter 400 kg/ha of 10-10-20	200 kg/ha of 10-10-20 after 15, 30, 50and 60 days of plantation
Chilli	15-20 t/ha of organic matter 300 kg/ha of 10-10-20	150 kg/ha of 10-10-20 two to four weeks later subculturing
Cassava	5-10 t/ha of organic matter 600 kg/ha of 10-10-20	-

Sources: CDH data sheets, 1989

Appendix 11: Solar drip irrigation technique

Solar drip is a highly efficient, simple and cost-effective irrigation system that can help millions of farmers around the world to increase production while up to 10 times less water.

The solar drip, also known as the "Kondenskompressor", is a watering technique designed achieve optimum use of water by using solar energy as the driving force behind the distillation and movement of the water. It's a surprisingly simple and effective system that can reduce the amount of water used for watering by around 10 times compared with traditional watering systems.

The Kondenskompressor is made from readily available PET plastic bottles. It is easy to make and install, and can be used by any farmer, whether private or professional. It requires minimal maintenance, i.e. adding water to the deposit when necessary and removing any plants that have grown inside the Kondenskompressor.

Using this technique allows plants to develop fully, using only the amount of water they need and avoiding evaporation of any surplus.

Because it uses readily available recycled materials and is extremely simple to manufacture and install, this technique can be easily used in poor countries with long dry seasons, and even in desert areas that could be supplied with fresh or salt water (e.g. from the sea).

How to make a Kondenskompressor and apply the solar drip technique

To make a kondenskompressor, all you need are two PET plastic bottles with caps (plastic water bottles) of different sizes (for example, one 5-litre and one 1.5-litre). Glass bottles with the same characteristics can be used if there is a means of cutting them.

The largest bottle should be cut at the base, while the smallest should be cut about halfway down and the bottom half used.

The bottom half of the small bottle, filled with water, will be placed on the floor and the large bottle will be placed on top. The relative position between the two should allow the cap of the large bottle to be opened to pour water into the small bottle.

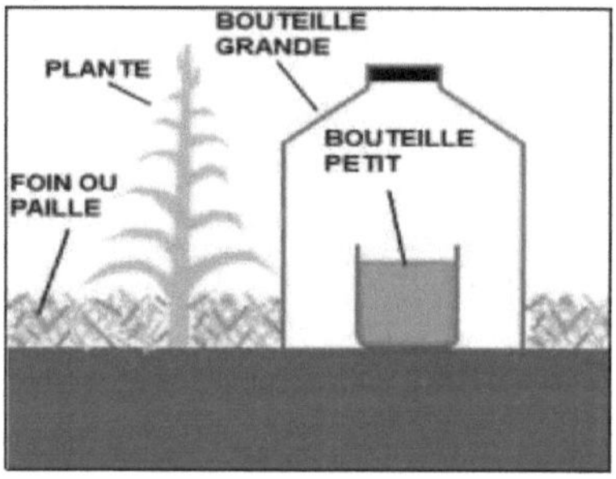

The two bottles (the kondenskompressor) are placed next to the plant to be watered. Hay, straw or dry leaves are placed around the plant and the kondenskompressor. When the sun's rays reach the kondenskompressor, they produce a greenhouse effect inside it, raising the air temperature and causing the water inside the tank to evaporate. The air inside the bell becomes saturated with moisture, causing condensation in the form of water droplets on the wall. As long as the kondenskompressor remains exposed to the sun, evaporation will continue, producing larger and larger drops of water that will eventually slide down the walls and fall to the ground to water it. In this way, a small-scale natural water cycle is created.

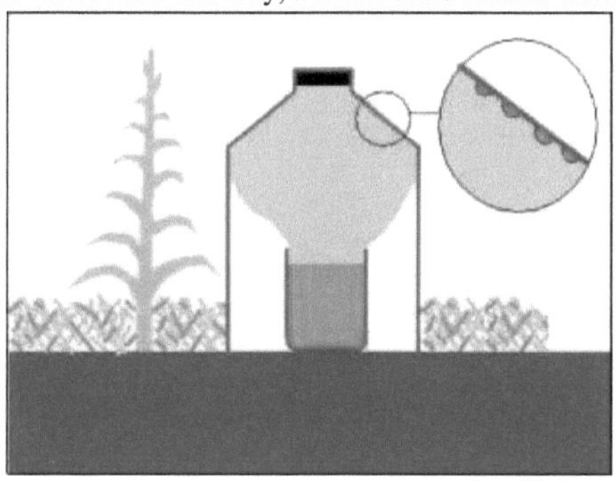

When the sun hits the kondenskompressor, the water evaporates inside the container and condenses on the walls of the large bottle.

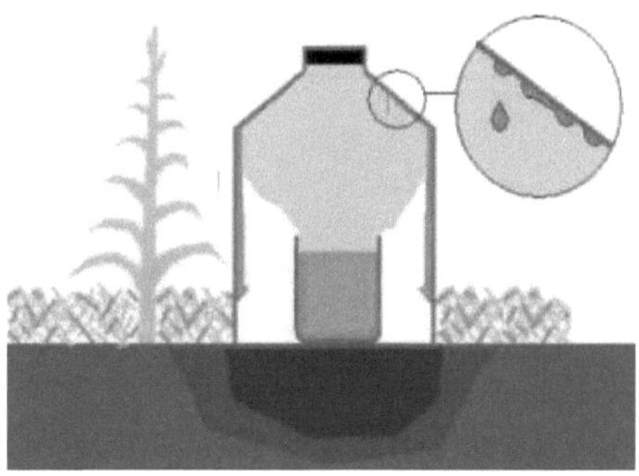

If the Kondenskompressor is left in the sun, evaporation will continue, the drops of water will increase and slide down the walls to the ground, where they will moisten it.
The advantage of these drip irrigation systems, both solar and conventional, is that the plant receives exactly the right amount of water needed for its development, avoiding wastage due to over-watering, evaporation or filtration.
As there is no direct connection with the outside atmosphere, no water is lost through the bell. To avoid the risk of evaporation, hay or straw is placed in the area between the plant and the Kondenskompressor to conserve soil moisture. In this way, water can only evaporate when it has been used by the plant for its development, through transpiration of the leaves.
We did not consider it necessary to include this method in the chapters of our study, because we felt it should first be tested to ensure its reliability. Hence the interest of this proposal, as it could be the subject of a more in-depth study.

Summary

The consequences of the salinisation of the Niayes on market gardening gave rise to a need that we felt following the environmental problems observed at the site. Analysis of the physical environment, with particular emphasis on the Niayes, will enable us to understand the causes of the various problems and may help the competent authorities to opt for better management of this highly vulnerable area. We started from the main hypothesis that "salinisation of the Niayes has a negative impact on market gardening". To verify this, we conducted a field survey, interviewing market gardeners to get their views on the current situation. The overview of the area shows an average level facilities and diversified activities, such as market gardening, arboriculture, etc. The anarchic occupation of the area also contributes to its degradation. This is compounded by the degradation of natural resources, not to mention the negative impact on the soil of the pesticides used by market gardeners. Heavy pressure on land is a major obstacle to the development of the site. Interpretation of the results of chemical analyses of the soil and water throughout the zone at our various sites has shown a rapid change in the level of salinity over the years, with a very alkaline ph, alkaline soil and saline water, unsuitable for irrigation and market gardening.

Key words

Niayes, salinisation, irrigation water, degradation, management, natural resources, market gardening, land pressure.

MIX
Papier aus verantwortungsvollen Quellen
Paper from responsible sources
FSC® C105338

FSC
www.fsc.org

Printed by Books on Demand GmbH, Norderstedt / Germany